India Studies in Business and Economics

India is one of the fastest growing economies of the world, with the country being an important G20 member. It is a significant player in the global market with a notable demographic advantage, and an important geo-political and geo-trade position. Additionally, due to its position of being not only the most populous country in the world, but also the world's largest youth population, India has a pivotal role to play in meeting global SDG goals. This makes research from the country to be of immense interest to the global community.

With this backdrop, this series aims to bring forth the latest studies and research specific to India from all areas of economics, finance, business, trade, and management science, often with strong social science and political economy linkages. The titles featured in this series present rigorous research, both empirical and non-empirical in nature, usually accompanied by policy recommendations. They evoke and evaluate various aspects of the economical and/or the business and management landscape in India, with a special focus on India's relationship with the world. The series also tracks research on India's position on social issues, on health, on politics, on agriculture, on rights, and many such topics which directly or indirectly affect sustainable growth of the country.

The series welcomes contributions from scholars or practitioners belonging to the academic community, corporates, think tanks and/or non-governmental organizations, and also policymakers to publish their work in the form of authored monographs, edited volumes or collection of case studies. Each proposal undergoes at least two double blind peer review where a detailed concept note along with extended chapter abstracts and/or a sample chapter is peer reviewed by experienced academics. The reviews may be more detailed if recommended by reviewers. The series follows the Ethics Statement as per Springer Nature's standard guidelines.

G C Manna

75 Years of the Indian National Sample Survey

Evolution of Sample Design, Key Challenges and Way Forward

INSTITUTE FOR HUMAN DEVELOPMENT
NEW DELHI • RANCHI
www.ihdindia.org

NCAER
SINCE 1956
QUALITY. RELEVANCE. IMPACT

Springer

G C Manna
Institute for Human Development
New Delhi, Delhi, India

National Council of Applied Economic
Research
New Delhi, Delhi, India

ISSN 2198-0012 ISSN 2198-0020 (electronic)
India Studies in Business and Economics
ISBN 978-981-96-7319-3 ISBN 978-981-96-7320-9 (eBook)
https://doi.org/10.1007/978-981-96-7320-9

This Springer imprint is published by the registered company Springer Nature Singapore Pte Ltd.
The registered company address is: 152 Beach Road, #21-01/04 Gateway East, Singapore 189721, Singapore

Foreword

Since its establishment in 1950, the National Sample Survey (NSS) has played a pivotal role in shaping India's socio-economic statistics landscape, significantly aiding the country's policymaking process. As one of the most scientifically structured sample survey systems among developing countries, India's NSS stands out for statistical excellence and it has earned global recognition.

Over the years, the NSS has expanded considerably, both in scope and frequency of data collection. The availability of crucial statistics—covering household consumption, employment, unemployment, enterprises and more—has not only enriched policymaking process but has also contributed to academic research and public discourse.

Dr. G C Manna, with his extensive career in India's statistical system, has made invaluable contributions to the official statistical framework, particularly the NSS. Having served as Director General of the NSSO and CSO and as a member of the National Statistical Commission, he has been instrumental in shaping statistical methodologies and practices. Even after his formal retirement from the Government of India, he remains actively engaged with various government committees and academic institutions. The Institute for Human Development (IHD) and the National Council of Applied Economic Research (NCAER) have been privileged to host Dr. Manna and benefit from his insights and experience.

As the NSS nears its 75th anniversary, there is a need for an authoritative account documenting its evolution and achievements. Dr. Manna, with his deep expertise, is uniquely suited for this task, and he has executed it admirably. This book offers a comprehensive overview of the NSS's journey, detailing its survey methodologies, sampling design, estimation procedures and changes over time. The book also provides insights from international experiences and thoughtful recommendations for strengthening the system for the future.

In today's world of increasing uncertainty, the significance of reliable and comprehensive timely statistics has never been greater. This book is a timely and important contribution, written in an accessible style, making it a valuable resource for researchers, policymakers, students and the statistical community at large. We extend our appreciation and best wishes to Dr. Manna and hope he continues making his valuable contributions to the field of statistics for many more years.

New Delhi
June 2025

Alakh N. Sharma
Professor and Director
Institute for Human Development
New Delhi, India

Anil K. Sharma
Secretary and Operations Director
National Council of Applied Economic Research
New Delhi, India

Preface

The Indian National Sample Survey (NSS) is one among the largest surveys of its kind in the world. The NSS was set up way back in 1950 primarily with the objective of conducting large-scale and nationwide sample surveys to fill up data gaps on different subjects of enquiry. This book is a modest attempt to give an overview of the survey design of the NSS with a summarization of the sample design adopted in the NSS since its inception till now. While certain documentations on the evolution of the sample design in the NSS are accessible, no single document is publicly available that gives a complete overview of the sample design adopted in various rounds of the NSS since its inception till today. The book aims to fill up this gap. Further, the lessons regarding the level of precision of the NSS estimates of some key indicators based on the recent surveys on different subjects of enquiry are also documented in the book with certain suggestions for possible remedial measures. A chapter on some global survey practices adopted by certain countries in respect of surveys on household consumption expenditure and labour force surveys is also included for cross-country comparison of the methodologies to draw lessons that might help in properly designing the surveys on the relevant subjects. The concluding chapter of the book highlights some of the key challenges and way forward that might require attention for further improvement of the database of the NSS.

For the relevant information in a substantial portion of the book, I depended upon various NSS reports and other documents accessed from the website of the Ministry of Statistics and Programme Implementation, Government of India for which I am thankful to the ministry for its initiatives to upload these documents on the ministry's website. I am also thankful to Ms. Nitika Gupta, Agricultural Statistics Wing, Faridabad, Field Operations Division, National Sample Survey Office, Government of India for her suggestions on the draft write-up relating to Sample Checks on Area Enumeration and Crop Estimation Surveys.

As regards the contents in Chapter 6 on global survey practices, I am thankful to the officers of the Australian Bureau of Statistics, Office for National Statistics (ONS), Statistics Canada and U.S. Bureau of Labor Statistics who interacted with me through email communications regarding their views/suggestions on the draft write-up that I had shared with them relating to the Household Expenditure Surveys

(Living Costs and Food Survey by the Office for National Statistics) and Labour Force Surveys in the respective countries. I am thankful to Dr. Denisha, Customer Advisor Strategic Communications, ONS for linking me with the relevant team of the ONS. I also thank Dr. Michael Bassett, Household Living Conditions Community, Wellbeing and Aboriginal and Torres Strait Islander Statistics Branch, People & Place Division, Australian Bureau of Statistics; Dr. Paul Bloomfield, Data Advice and Relations Team, Social Survey Transformation, Directorate of Surveys, Office for National Statistics; and Dr. Jeannette Rodillo, Consulting-Analyst, Statistics Canada for confirming the correctness of the draft write-up on the related surveys in the respective countries that I shared with them.

I acknowledge with thanks the comments and suggestions on the draft manuscript of the book received from Professor Sudipto Mundle (Chairman, Centre for Development Studies, Thiruvananthapuram), Professor Alakh N. Sharma (Director, Institute for Human Development (IHD), New Delhi) and Professor Ravi Srivastava (Professor, IHD). I am particularly thankful to Professor Alakh N. Sharma, Director (IHD) for his generous support in the release of this book. I also thank Ms. Priyanka Tyagi (CEO, IHD), Ms. Jyoti Girish (Administrative Officer, IHD) and Ms. Varsha Susan Mathew (Research Assistant, IHD) for their support in the finalization of the manuscript. Ms. Deepika Chaddha is appreciated for her editorial support. I thank my colleagues in the National Council of Applied Economic Research—particularly Professor Anil K Sharma, Professor Poonam Munjal and Professor Bornali Bhandari—for their encouragement in bringing out this publication. I am also thankful to the Springer team, especially Ms. Nupoor Singh, for their assistance in publishing the book within a short time. Finally, I thank my wife, Manju, for the support she provided to me in writing this book.

It is needless to mention that the views expressed in this book are my personal views and not of the organizations to which I belong. I took utmost care to see that the contents in this book correctly report/reproduce the various facts and figures accessed from different sources including the NSS. However, I apologize for inadvertent error, if any, while reporting these facts. It is a matter of great satisfaction that the year of publication of this book coincides with the celebration year of 75 years of the NSS. I hope that the readers will find this book useful.

New Delhi
June 2025

G C Manna

Contents

Abbreviations

ABS	Australian Bureau of Statistics
ALE	Agricultural Labour Enquiry
ASI	Annual Survey of Industries
ASSSE	Annual Survey of Services Sector Enterprises
ASUSE	Annual Survey of Unincorporated Sector Enterprises
AYUSH	Ayurveda, Yoga & Naturopathy, Unani, Siddha, Sowa Rigpa and Homoeopathy
BA	Bazaar Area
BLS	Bureau of Labor Statistics
BR	Business Register
BRR	Balanced Repeated Replication
CAMS	Comprehensive Annual Modular Survey
CBSA	Core-Based Statistical Areas
CCE	Crop Cutting Experiment
CE	Consumer Expenditure Surveys
CEB	Census Enumeration Block
CEO	Chief Executive Officer
CMA	Census Metropolitan Area
CMP	Composite
CMS	Comprehensive Modular Survey
CPD	Coordination and Publication Division
CPI	Consumer Price Index
CPIAL	Consumer Price Index for Agricultural Labourers
CPIRL	Consumer Price Index for Rural Labourers
CPS	Current Population Survey
CSO	Central Statistics Office / Central Statistical Office
CWS	Current Weekly Status
DE	Directory Establishment
DU	Dwelling Unit
DESs	Directorates of Economics and Statistics
DG	Director General

DIID	Data Informatics & Innovation Division
DME	Directory Manufacturing Establishment
DNH & DD	Dadra and Nagar Haveli & Daman and Diu
DPD	Data Processing Division
DPIIT	Department for Promotion of Industry and Internal Trade
DTE	Directory Trading Establishment
EA	Enumeration Area
EB	Enumeration Block
EC	Economic Census
EPFO	Employees' Provident Fund Organisation
ESIC	Employees' State Insurance Corporation
E–U	Employment–Unemployment
EUS	Employment–Unemployment Survey
FOD	Field Operations Division
FSU	First-stage Unit
GCES	General Crop Estimation Surveys
GDP	Gross Domestic Product
GVA	Gross Value Added
GVAPW	Gross Value Added Per Worker
HA	Hospital Area
HCE	Household Consumer/Consumption Expenditure
HCES	Household Consumption Expenditure Survey
HES	Household Expenditure Survey
HG	Hamlet-Group
HH	Household
HS	Higher Secondary
IA	Industrial Area
ICLS	International Conference of Labour Statisticians
ICP	International Comparison Programme
ICS	Improvement of Crop Statistics
IES	Income and Expenditure Survey
IHDS	India Human Development Survey
ISI	Indian Statistical Institute
IS Wing	Industrial Statistics Wing
IV Units	Investigator Units
LB	Labour Bureau
LCF	Living Costs and Food Survey
LFPR	Labour Force Participation Rate
LFS	Labour Force Survey
LLH	Land and Livestock Holding
LUS	Land Utilization Survey
MAF	Master Address File
MCA	Ministry of Corporate Affairs
ML	Means of Livelihood
MMRP	Modified Mixed Reference Period

MoSPI	Ministry of Statistics and Programme Implementation
MPCE	Monthly Per Capita (Consumer) Expenditure
MRP	Mixed Reference Period
MSE	Mean Square Error
NAD	National Accounts Division
NAR	Net Attendance Ratio
NAS	National Accounts Statistics
NCAER	National Council of Applied Economic Research
NCR	National Capital Region
NDE	Non-Directory Establishment
NDME	Non-Directory Manufacturing Establishment
NDTE	Non-Directory Trading Establishment
NIC	National Industrial Classification
NSC	National Statistical Commission
NSO	National Statistical Office
NSS	National Sample Survey
NSSO	National Sample Survey Organisation/National Sample Survey Office
OAE	Own-Account Enterprise
OAME	Own-Account Manufacturing Enterprise
OATE	Own-Account Trading Enterprise
OWS	Occupational Wage Survey
PCA	Primary Census Abstract
PLFS	Periodic Labour Force Survey
PPA	Probability Proportional to Area
PPS	Probability Proportional to Size
PPSWR	Probability Proportional to Size with Replacement
PSU	Primary Sampling Unit
QLFS	Quarterly Labour Force Survey
RO	Regional Office
RSE	Relative Standard Error
SA	Slum Area
SANKALP	Skill Acquisition and Knowledge Awareness for Livelihood Promotion
SAS	Situation Assessment Survey
SASAs	State Agricultural Statistics Authorities
SB	Sub-Block
SC	Scheduled Caste
SDRD	Survey Design and Research Division
SHS	Survey of Household Spending
SIH	Survey of Income and Housing
SNA	System of National Accounts
SRO	Sub-Regional Office
SRSWOR	Simple Random Sampling without Replacement
SS	Second-stage

SSI	Small Scale Industries
SSS	Second-stage Stratum/Strata
ST	Scheduled Tribe
SU	Sub-unit
TUS	Time Use Survey
UFS	Urban Frame Survey
UK	United Kingdom
UPR	Usual Place of Residence
UPSS	Usual Principal and Subsidiary Status
UR	Unemployment Rate
URP	Uniform Reference Period
U.S.	United States
USU	Ultimate-stage Unit
UT	Union Territory
WPI	Wholesale Price Index
WPR	Worker Population Ratio
ZO	Zonal Office

Chapter 1
Introduction

1.1 Background and Chapterization

The Indian National Sample Survey (NSS) was set up way back in the year 1950 with a view to filling up the data gaps on various subjects of the Indian economy by conducting nationwide sample surveys on a scientific basis. The formative years of the NSS faced quite a few challenges in the development of the survey design in view of (a) the limited availability of the information to be satisfactorily used as the sampling frame for the selection of villages and urban blocks in the household and establishment surveys, (b) constraint of adequate manpower for carrying out the fieldwork, and (c) other survey related issues including the problems of coordination between the Indian Statistical Institute (ISI)—which was responsible for designing the survey, processing of data and preparation of reports based on the NSS—and the Directorate of National Sample Survey in the Government of India which handled the aspect of data collection through its field staff. Subsequently, with the set-up of the National Sample Survey Organisation (NSSO) which became fully operational by the middle of the year 1972, all the survey related activities of the NSS taken care of by the NSSO got adequately strengthened.

Over almost the past 75 years, the sample design of the NSS has witnessed a number of significant changes and refinements depending upon the subjects of enquiry in its various surveys. For example, the recently introduced (i.e., since 2017–18) high-frequency Periodic Labour Force Survey (PLFS) has modified the traditional sample design by way of adoption of the rotational sampling design for selection of sample blocks in the urban areas to begin with and thereafter the procedure extended to rural areas also in PLFS: 2025. In the past, the NSS also conducted certain pilot studies and it gained useful insights in the process. The experiences in terms of the level of precision of the NSS estimates at the State/Union Territory level are also mixed depending upon the subjects of enquiry. There is no single document which gives a comprehensive picture covering all these aspects of the NSS. This book is a modest attempt to fill up the said gap. It has been written keeping in view that

G C Manna, *75 Years of the Indian National Sample Survey*, India Studies in Business and Economics, https://doi.org/10.1007/978-981-96-7320-9_1

different types of readers—including the students/researchers, other academics, data users of the NSS and even the data producers—may be interested in referring the book.

The book contains nine chapters. Apart from the above background, Chapter 1, after briefly describing the structure of statistical system in India, also highlights the rationale behind formation of the NSS and the NSSO including the major activities of the NSSO. The conduct of socio-economic enquiries is the most important activity of the NSSO. The details about these surveys including the sample design form the core of discussion of this book as part of chapters 2 to 9. However, Chapter 1 discusses about the salient features of the three other important activities of the NSSO namely the conduct of Annual Survey of Industries, Urban Frame Survey and Sample checks on area enumeration and crop estimation surveys.

For those who may like to conduct a primary survey in a geographical region or for any particular domain of interest, the salient features of the survey methodology of the NSS as summarized in Chapter 2 may help in designing the survey. Some of the important features of the NSS, which are discussed in this chapter, include the sampling frame of first-stage units (FSUs) in the NSS, methods of stratification and sampling of FSUs along with a summary of methods of sampling of FSUs used in different NSS rounds starting from the first round till today, and the concept of NSS region as well as the complete list of such regions. For the benefit of the readers including the data users, various subjects covered in the different rounds of the NSS till today are also described in this chapter.

With illustrations from specific rounds covering different subjects of enquiry, Chapter 3 gives the glimpses of sample design adopted in these surveys after dividing the entire time period of the NSS till today into four phases. This chapter is likely to give a fair idea about how the sample design in the NSS evolved over time. These four phases include (i) the initial phase from the year 1950 (being the year when the NSS was launched) till mid 1972 i.e., the time since when the NSSO started its functioning in full form with the Governing Council overseeing all technical aspects of the surveys conducted by the organization; (ii) the second phase from 1972 till 1999–2000 which coincides with the survey period of NSS 55th round that took up the subjects of household consumption expenditure and employment-unemployment with certain significant changes introduced in the reference period of data collection relating to household consumption expenditure; (iii) the third phase from 2000–01 till the year 2016–17; and finally (iv) the fourth/last phase since 2017–18 which is a period with some high-frequency surveys introduced. The first few rounds of the NSS in its initial phase witnessed certain major changes in the sample design and method of stratification for the selection of villages and blocks in view of limited availability of the information to be satisfactorily used as the sampling frame for selection of samples.

Chapter 4 highlights some of the most significant changes in the sample design since the inception of the NSS till today. Chapter 5 describes three pilot surveys conducted by the NSSO namely Pilot Survey of Income, Consumption and Savings (September 1983–December 1984); Study on Alternative Approach of Data Collection for Measuring Gross Value Added through Enterprise Surveys (Mid-nineties);

and Pilot Survey on Suitability of Different Reference Periods for Measuring Household Consumption (January–June 2000) including the major findings from these surveys.

Chapter 6 describes global survey practices followed by certain countries (namely, Australia, Canada, South Africa, the United Kingdom and the United States apart from India) in respect of surveys on household consumption expenditure and labour force surveys. The chapter also summarizes the major differences in the approach adopted by India with those of other countries with some way forward. Many students and researchers are likely to be interested in undertaking some tabulation exercises based on the unit level data available from the surveys. For their benefits, Chapter 7 explains the concept of 'multiplier' as well as the method of estimation of aggregates and ratios based on the NSS data with some elaborations as to how actually such estimates are to be derived from the unit level data.

The survey results in any sample survey are subject to margins of error in the sense that the estimates based on the survey are likely to deviate from the true values in the population by certain magnitudes. Chapter 8 is devoted to the discussion on the magnitudes of error associated with some key estimates based on the NSS. This is likely to help the data users analyze the survey estimates in proper perspectives. Further, the lessons drawn regarding the margins of error as per the present survey methodology along with the suggestions given might even help the data producers to refine the survey instruments in the forthcoming surveys. Finally, Chapter 9 highlights certain challenges which the NSS may perhaps address to further improve its database. Some of the critical areas in this context which have been highlighted for attention include the issue of underestimation of population in the NSS and the major issues with the surveys on household consumption expenditure, PLFS and the Enterprise Survey.

1.2 Statistical System in India

The development of Indian Statistical System has a long history. However, it has been mainly a post-independence phenomenon. The statistical system in India functions within the overall administrative framework of the country. The division of responsibility for collection of statistics on different subjects between the Central Government and State Governments is based on the classification of subjects under the Union List, the State List and the Concurrent List as detailed in the Indian Constitution. Nationwide 'large-scale' statistical activities like Population Census, Agricultural Census, Livestock Census, Economic Census and the countrywide sample surveys as well as compilation of macro-economic aggregates like national accounts and All-India Price Indices are mainly Central activities with substantial involvement of State agencies in data collection. The State Governments also collect data on a number of variables of interest to the States.

Within the Central Government, the allocation of subjects among various ministries and departments is as per the Allocation of Business Rules, 1961 and subsequent amendments which take place from time to time. The respective States also have a statistical system similar to that in the Centre and the system is decentralized laterally over various departments of the State Government.

Most of the Ministries in the Central Government have either a full-fledged statistical department or a section dealing with the collation of statistics relevant for the Ministry. The responsibility of overall coordination of all statistical activities at the all-India level vests with the Central Statistical Office (CSO) in the Ministry of Statistics and Programme Implementation (MoSPI), Government of India.[1] Similarly, state-level coordination of statistical activities among different departments of the State Government is performed by the respective State Directorates of Economics and Statistics (DESs). While doing so, the State DESs keep liaison with the CSO for coordination at the all-India level. Generally, the State DESs have larger establishments at the headquarters with statistical offices spread in the districts or at least in the regions of the State. So far as the statistical personnel in India are concerned, a majority of them work in the State Governments.[2]

1.3 Creation of the National Sample Survey

The absence of adequate statistics on various aspects of the economy had been felt for a long time. After India achieved her independence in 1947, the government realized this deficiency. In fact, the Government of India Resolution No. 15(33)-P/49 dated the 4th August 1949, appointing the National Income Committee under the chairmanship of Professor P.C. Mahalanobis and D.R. Gadgil and V.K.R.V. Rao as members to prepare a report on the National Income and related estimates, also makes a mention about the data gaps. To quote: 'The Government of India have been giving consideration for some time to the inadequacy of the factual data available for the formulation of economic policies'.[3] Subsequently, while preparing the estimates of national income for the years 1948–49, 1949–50 and 1950–51, the National Income Committee found large gaps in statistical data required for compilation of national income statistics, especially those relating to the unorganized sector of the economy. Some of the important areas/subjects with inadequate data as highlighted in the report

[1] The MoSPI has two wings: one relating to Statistics and the other Programme Implementation. The Statistics Wing called the National Statistical Office (NSO) comprises the Central Statistical Office (CSO), the Data Informatics & Innovation Division (DIID) and the National Sample Survey Office (NSSO). The CSO, established in May 1951, was earlier named as Central Statistical Organisation.

[2] As per the status prevailing during 2009, while 34,681 statistical personnel (with relative shares of the State DESs and Line Departments being 21,819 and 12,862) were employed by the State Governments (considering all the three categories of personnel namely Professional, Intermediate and Primary), the Central Ministries/Departments/Organizations employed 9,929 such persons [MoSPI (2009)].

[3] Government of India (1954). Chapter 1, Introduction.

of the committee[4] include crop area statistics; estimation of yield of various crops; costs of cultivation; livestock numbers and yields of many kinds of animal products; small scale industry; and employment and income data of trade sector as well as transport sector other than railways.

With the above realization and accordingly as per the advice of Professor P.C. Mahalanobis, then Honorary Statistical Adviser to the Cabinet, the scheme to set up the Indian NSS was approved by the Government in the year 1950 with the active support of the late Pandit Jawaharlal Nehru, the Prime Minister of India at that time.[5]

1.4 Establishment of the National Sample Survey Organisation

The NSS was introduced primarily with the objective of filling up data gaps in various socio-economic characteristics through continuing and countrywide sample surveys. The Directorate of National Sample Survey was set up in the year 1950 in the Ministry of Finance, Government of India to collect data on various socio-economic subjects on a continuing basis as recommended by the National Income Committee. In the formative years after the creation of the NSS, the ISI handled most of the work relating to the NSS including the formulation of sample design, preparing the survey questionnaires, training of field personnel, data processing as well as preparation of survey reports. However, the fieldwork for various surveys was carried out by the Directorate of NSS in all the areas of the country except West Bengal and the city of Bombay where the fieldwork was conducted by the ISI. This division of responsibility between the ISI and the Directorate of NSS had lot of coordination issues to complete the survey work and prepare the reports within a reasonable time schedule. To obviate this difficulty and for ensuring a better coordination and effective programming of various survey activities, as per the recommendation of a three-man committee whose members were B. Sivaraman (Chairman), V.M. Dandekar and R.R. Bahadur, the work of the NSS was reorganized by shifting the design and analysis wing of the ISI to the Department of Statistics, Government of India and establishing the National Sample Survey Organisation (NSSO) in January 1971, with all activities related to the NSS brought under the control of the NSSO.[6]

[4] For details, refer to Recommendations on the Improvement of Data, in Chapter Three of the Final report of the committee.

[5] The NSS was started by the Ministry of Finance, Government of India.

[6] The NSSO was established under the Department of Statistics, Government of India as per a resolution dated 5 March, 1970. The name 'National Sample Survey Organisation' was subsequently changed to 'National Sample Survey Office'. As of now, the NSSO is a part of the Statistics Wing, called the National Statistical Office (NSO), of the MoSPI, Government of India which came into existence as an independent Ministry on 15.10.1999 after the merger of the then Department of Statistics and the Department of Programme Implementation [Source: MoSPI (2025c)].

The NSSO started functioning with a Governing Council at its apex having full autonomy in the matter of collection and processing of data including the publication of NSS reports without any undue interference. V.M. Dandekar was the first Chairman of the Governing Council, NSSO. Apart from the non-official Chairman, the Governing Council used to have five academicians, five data users from Central and State Government departments and senior officers of the Department of Statistics/MoSPI as the members with the Director General and Chief Executive Officer (DG & CEO) of NSSO as the Member Secretary of the Council. The Governing Council of NSSO has since been abolished and its activities are now overseen by the National Statistical Commission (NSC).[7]

1.5 Organizational Structure of the NSSO

The National Sample Survey Organisation, of late renamed as the National Sample Survey Office, is headed by a Director General. Any large-scale sample survey, like the socio-economic surveys conducted by the NSSO as part of the various NSS 'rounds', primarily involves four major activities viz. (i) formulation of survey methodology including preparation of sample design, development of questionnaire as well as instructions manual, data edit programmes and drafting of the survey report; (ii) execution of the fieldwork; (iii) processing of primary data collected through surveys; and finally (iv) the work of coordination of the entire survey work for ensuring timely release of data. To handle these activities smoothly, NSSO started operating with its four divisions respectively taking care of these four activities: Survey Design and Research Division (SDRD), Field Operations Division (FOD), Data Processing Division (DPD), and Coordination and Publication Division (CPD). The names of the four divisions have undergone changes over time. Currently, they are respectively known as Household Survey Division, Field Operations Division (name unchanged), Enterprise Survey Division and Coordination and Quality Control Division.

The Household Survey Division with its Headquarter in Kolkata has also two units namely, Household Survey Unit—Bengaluru and Household Survey Unit—Nagpur. The Enterprise Survey Division has headquarters in Kolkata and it has three units namely, Enterprise Survey Unit—Ahmedabad, Enterprise Survey Unit—Delhi and Enterprise Survey Unit—Giridih. Among the four divisions of NSSO, FOD is the largest one with its headquarters located in Delhi. Agricultural Statistics Wing of FOD Headquarters is located at Faridabad. In addition, FOD has 6 Zonal Offices (at Bengaluru, Guwahati, Jaipur, Kolkata, Lucknow and Mumbai), 53 Regional Offices

[7] As per the recommendation of the Rangarajan Commission, which reviewed the Indian Statistical System and submitted its report in 2001, and with the subsequent approval of the Cabinet, the NSC was constituted with effect from 12th July 2006 with a mandate to evolve policies, priorities and standards in statistical matters. Apart from a Chairman, the NSC has four Members. CEO, NITI Aayog is an Ex-Officio Member of the NSC and Chief Statistician of India & Secretary, MoSPI is the Member Secretary of the NSC.

Statement 1.1 Distribution of Zonal Offices (ZO), Regional Offices (RO) and Sub-Regional Offices (SRO) of FOD by State and Union Territory

State/UT	Number of Offices			State/UT	Number of Offices		
	ZO	RO	SRO		ZO	RO	SRO
Andaman & Nicobar Is		1	–	Madhya Pradesh		3	8
Andhra Pradesh		2	6	Maharashtra	1	4	8
Arunachal Pradesh		1	–	Manipur		1	–
Assam	1	2	3	Meghalaya		1	1
Bihar		2	5	Mizoram		1	–
Chandigarh		1	–	Nagaland		1	–
Chhattisgarh		1	3	Odisha		2	4
Delhi*		1	–	Puducherry		–	1
Goa		1	–	Punjab		2	6
Gujarat		2	8	Rajasthan	1	2	5
Haryana**		–	4	Sikkim		1	–
Himachal Pradesh		1	3	Tamil Nadu		3	8
Jammu & Kashmir		2	3	Telangana		1	3
Jharkhand		1	5	Tripura		1	–
Karnataka	1	2	6	Uttar Pradesh	1	4	13
Kerala		2	6	Uttarakhand		1	1
				West Bengal	1	3	7
				India	6	53	117

Source MoSPI (2025d)
*Apart from one RO, FOD has headquarter in Delhi
**In addition to the four SROs, Agricultural Statistics Wing of FOD headquarter is also located in the State at Faridabad

and 117 Sub-Regional Offices with the State-wise distribution shown in Statement 1.1. The Coordination and Quality Control Division is located in Delhi. (MoSPI, 2025d).

1.6 Activities of the NSSO

1.6.1 Socio-Economic Surveys and Other Allied Activities

The most important activity of the NSSO has been to conduct socio-economic surveys on a wide range of subjects in the form of successive '*NSS Rounds*'. The NSSO is also involved in the collection of price data namely (a) monthly rural retail price data for 260 items from 603 villages spread across 26 States/UTs for compilation of

Consumer Price Index (CPI) for Agricultural Labourers (CPIAL) and CPI for Rural Labourers (CPIRL) by the Labour Bureau with the current Base Year as 1986–87 for both the indices; (b) daily wage rates for 12 major agricultural occupations[8] and 13 major non-agricultural occupations[9] from the same set of villages; (c) monthly price data from the markets located in 1181 villages and 310 towns (with 1078 quotations per month) across the country for construction of CPI (Rural) and CPI (Urban) by the CSO with the existing Base Year of the index as 2012; and (d) monthly price data from about 6000 factories in the organized manufacturing sector which are used in the compilation of Wholesale Price Index (WPI) with current Base Year as 2011–12 by the Department for Promotion of Industry and Internal Trade (DPIIT), Ministry of Commerce & Industry, Government of India. These apart, FOD of NSSO also collects price data under the International Comparison Programme (ICP).

In addition, the NSSO carries out the Annual Survey of Industries (ASI),[10] conducts Urban Frame Survey (UFS) to prepare urban sampling frames for selecting area samples (i.e., urban blocks) of the socio-economic surveys and performs sample checks on area enumeration and crop estimation surveys done by the States. While Chapter 2 and subsequent chapters of the book mainly concentrate on the socio-economic surveys (other than ASI) conducted by the NSS, subsequent text/paragraphs in this chapter describe briefly some of the important aspects of the ASI, UFS and the work related to Sample checks on area enumeration and crop estimation surveys for the benefit of the readers.

1.6.2 Annual Survey of Industries

1.6.2.1 The Coverage

The Annual Survey of Industries (ASI) is one of the principal sources of industrial statistics in India. It has been conducted since 1959 under the Collection of Statistics Act, 1953. Presently, the survey is being conducted under the Collection of Statistics Act, 2008. The Survey, inter-alia, is aimed at estimating the size of the registered manufacturing sector in terms of number of establishments/factories and number of persons engaged by such establishments as well as the share of the

[8] Major agricultural operations include Ploughing, Sowing, Weeding, Transplanting, Harvesting, Winnowing, Threshing, Picking, Herdsman, Well digging and Cane crushing.

[9] Non-agricultural occupations include those of Carpenter, Blacksmith, Cobbler, Mason, Tractor driver, Sweeper and Unskilled labourers.

[10] Till recent past, the formulation of sample design of ASI, preparation of the questionnaire and writing of report based on ASI was the responsibility of the Industrial Statistics (IS) Wing which was under the CSO while the FOD of NSSO was responsible for carrying out the fieldwork of ASI. However, the IS Wing has since been brought under the administrative control of the Enterprise Survey Division of NSSO.

registered manufacturing sector in the overall manufacturing GDP.[11] The time series data generated by the ASI is useful in studying the changes in the structure of the registered manufacturing sector over time for formulation of industrial policies.

The ASI primarily covers manufacturing units[12] (called factories) registered under Section 2(m)(i) and 2(m)(ii) of the Factories Act, 1948 employing 10 or more workers using power or 20 or more workers not using power. However, since ASI 2014–15, the coverage of ASI has been extended to also cover the units with 100 or more employees which are not registered under Section 2 m (i) and 2 m (ii) of the Factories Act, 1948 but registered under any of the seven specified Acts/Authorities. These include (i) Companies Act, 1956; (ii) Shops and Commercial Establishment Act; (iii) Societies Registration Act; (iv) Factories Act. 1948; (v) Cooperative Societies Act; (vi) Khadi and Village Industries Board; and (vii) Directorate of Industries (District Industries Centre) in the Business Register of Establishments prepared by the State Governments. In addition, the ASI covers (a) bidi & cigar units, employing 10 or more workers with the aid of power and 20 or more workers without the aid of power which are registered under the Bidi & Cigar Workers (Conditions of Employment) Act, 1966 and (b) all electricity undertakings engaged in generation, transmission and distribution of electricity, not registered with the Central Electricity Authority. For the collection of data on most of the items/variables, the survey adopts the accounting year of the factory as the reference period. For example, in ASI 2022–23 the reference period was the accounting year of the factory ending on any day during the financial year: 2022–23.

1.6.2.2 The Present Composition of Census Scheme & Sample Scheme in ASI and the Sampling Procedure Since ASI 2015–16

For the collection of data on most of the variables including raw materials consumed, products and by-products manufactured and other expenses and receipts of the factories, reference period in ASI is the last accounting year of the factories with the closing date falling within the financial year under reference. ASI divides the universe i.e., the sampling frame of factories into two categories namely 'Census Scheme' and 'Sample Scheme' for sampling purposes. As the terminology suggests, all the factories as per the sampling frame appearing in the census scheme are surveyed on a

---

[11] However, in the current series of National Accounts Statistics (NAS) with 2011–12 as the base year, the data as per the ASI have somewhat limited use in the assessment of the share of the organized manufacturing sector in the GDP with dependence, in the current series, on the data on private companies created under the e-governance project, called MCA21 by the Ministry of Corporate Affairs. For further details in this regard, the readers may refer to MoSPI (2015).

[12] Apart from manufacturing activities, other industrial categories covered under ASI (2022–23) include (i) Cotton ginning, cleaning and bailing (National Industrial Classification i.e., NIC Code 01,632 as per NIC 2008); (ii) Seed processing for propagation (NIC Code 01,640); (iii) Salt production by evaporation of sea water or other saline waters (NIC Code 08,932); (iv) Repair and installation of machinery and equipment (NIC Code 33); (v) Waste collection, treatment & disposal activities (NIC Code 38); and (vi) Publishing activities (NIC Code 58). See MoSPI (2008a).

complete enumeration basis. As per the current sample design implemented since ASI 2015–16, the Census Scheme comprises the following:

(a) All industrial units belonging to less industrially developed States/UTs[13];
(b) All units engaged in salt extraction;
(c) Larger factories/units in terms of employment[14] in respect of the States/UTs other than those mentioned under (a);
(d) All units covered under Joint Return[15]; and
(e) The units in the 'strata' (with each stratum being State × District × Sector[16] × 3-digit of National industrial Classification i.e., NIC-2008) having 4 or less units after excluding the census sector units as defined under (a) to (d) above.

The Sample Scheme comprises all the remaining units in the frame not included in the Census Scheme. For the Sample Scheme in all the states, strata are formed by considering the factories under each State x District x Sector × 3-digit of NIC-2008. Before sampling of units from various strata with each stratum containing more than 4 units, the units in each stratum are first arranged in descending order of their number of employees. Thereafter, the required number of sample units is drawn based on circular systematic sampling (with equal probability).[17] An even number of units with a minimum of 4 units are selected from each stratum and they are distributed equally into four sub-samples. Out of these 4 sub-samples, two pre-assigned sub-samples are allotted to FOD for data collection (with corresponding samples including the units in the Census Scheme being treated as 'central sample') and the other two sub-samples are assigned to the State/UT participating in the ASI.

1.6.2.3 Census Scheme, Sample Scheme and Method of Stratification Adopted Earlier

Statement 1.2 gives the composition of census scheme, sample scheme and the method of stratification adopted earlier in the ASI.

[13] These include seven States/UTs of Arunachal Pradesh, Manipur, Meghalaya, Nagaland, Sikkim, Tripura and Andaman & Nicobar Islands from ASI 2015–16 to ASI 2018–19; and nine States/UTs i.e., including Mizoram and Ladakh also from ASI 2019–20.

[14] Larger units mean those having 50 or more employees in three States/UTs (Chandigarh, Delhi and Puducherry), 75 or more employees in six States/UTs (Bihar, Chhattisgarh, Himachal Pradesh, Jammu & Kashmir, Rajasthan and Kerala) and 100 or more employees in the remaining States/UTs.

[15] Furnishing of Joint Return is allowed when two or more units are located in the same State/UT belonging to the same industry (3-digit level of National Industrial Classification) under the same management.

[16] 'Sector' is broad economic activity group namely (i) Bidi, (ii) Manufacturing and (iii) Electricity.

[17] Such a method of sampling ensures representation of factories of different employment sizes within the overall sample of each stratum.

Statement 1.2 Composition of Census Scheme, Sample Scheme and Method of Stratification as Adopted Earlier in the ASI Prior to ASI 2015–16

Survey Period	Census Scheme	Sample Scheme / Stratum Composition
ASI 2014–15 (MoSPI, 2017)	Seven less industrially developed States/ UTs (as in ASI 2015–16); Rest of the States/ UTs: (i) units having 100 or more employees*, and (ii) all factories covered under Joint Returns; and Strata (State x District x Sector × 4 digit** of NIC-2008) having less than or equal to 4 units	Residual units (i.e., other than those in the census scheme) with each stratum formed on the basis of State x District x Sector × 4-digit** of NIC-2008 having more than 4 units in each stratum
ASI 2012–13 & ASI 2013–14 (MoSPI, 2016a)	Six less industrially developed States/ UTs (i.e., excluding Arunachal Pradesh); and others as in ASI 2014–15	As in ASI 2014–15
ASI 2009–10 to 2011–12 (MoSPI, 2013, 2016b, 2012)	Six less industrially developed States/ UTs (as in ASI 2012–13 & 2013–14); Rest of the States/ UTs: (i) units having 100 or more employees, and (ii) all factories covered under Joint Returns; and Strata (State@ x Sector × 4 digit of NIC-2008) having less than or equal to 4 units	Residual units with each stratum being composed of State@ x Sector × 4 digit of NIC-2008
ASI 2008–09 (MoSPI, 2011)	As above except that five less industrially developed States/UTs of Manipur, Meghalaya, Nagaland, Tripura and Andaman & Nicobar Islands were considered	As in ASI 2009–10
ASI 2004–05 to ASI 2007–08 (MoSPI, 2008b)	As in ASI 2008–09	As above except that State × 4-digit of NIC 2004 was considered for stratification
ASI 1997–98 to ASI 2003–04 (MoSPI, 2008b)	Twelve less industrially developed States/UTs[18] and for other States/UTs all the units having 200 or more workers as well as some significant units identified from the databases of ASI 1993–94 to ASI 1995–96 which had significant output	Rest of the units

(continued)

[18] These included A&N Islands, Chandigarh, Daman & Diu, Dadra & Nagar Haveli, Goa, Himachal Pradesh, Jammu & Kashmir, Manipur, Meghalaya, Nagaland, Pondicherry and Tripura.

Statement 1.2 (continued)

Survey Period	Census Scheme	Sample Scheme / Stratum Composition
ASI 1987–88 to ASI 1996–97 (MoSPI, 2008b)	Twelve less industrially developed States/UTs and for other States/UTs all the units with 100 or more workers irrespective of the use of power	Rest of the units with State X 3-digit industry group as the stratum with a plan to cover all the sample sector units in a span of three years
Till ASI 1986–87 (MoSPI, 2008b)	Twelve less industrially developed States/UTs and for other States/UTs all the units with 50 or more workers operating with power and units having 100 or more workers operating without power	Residual units employing less than 50/ 100 workers (operating with / without power) in the major States which were covered fully over a span of two years

Notes *The uniform threshold of 100 employees to define census scheme units has subsequently been replaced with a varying threshold of 50 / 75 / 100 employees for different States/UTs from ASI 2015–16 keeping in view the variability in the State-level composition of the large units by employee size
**Estimates at the district x NIC 4-digit level were generally found to have high relative standard errors and accordingly, the strata in ASI 2015–16 and onwards are being formed by considering NIC 3-digit in place of NIC 4-digit
@Deep stratification with district in place of State introduced during ASI 2012–13 to have provisions for the generation of district-wise estimates
Source Various MoSPI documents as mentioned in the table

1.6.2.4 The Current Sample Size

In ASI 2022–23, samples have been drawn considering an overall sampling fraction of 8% from all strata.[19] As per the same source, the size of the sampling frame for ASI 2022–23 was 255,244 factories of which 62,778 units belonged to the Census Scheme while the remaining 192,466 units comprised the Sample Scheme. Total allotted sample size for ASI 2022–23 was 82,734 units (62,778 in the Census Scheme and 19,956 in the Sample Scheme). However, a sample of 79,543 units could be actually surveyed. Out of 255,244 factories forming part of the sampling frame of ASI 2022–23, the estimated number of operating units in the country[20] during the year was 206,523. Statement 1.3 shows state-wise number of units in the frame of ASI 2022–23, number surveyed and the estimated number of operating units in the respective States and UTs.

---

[19] MoSPI (2024a).

[20] MoSPI (2024b).

Statement 1.3 Number of units/factories in the sampling frame, number of units surveyed and estimated number of units in operation by state and union territory ASI 2022–23

State/UT	Number of units			State/UT	Number of units		
	In the frame	Surveyed	Estimated number in operation		In the frame	Surveyed	Estimated number in operation
A&N Is	14	14	13	Madhya Pr	5,143	2,142	4,529
Andhra Pr	16,552	3,398	12,247	Maharashtra	26,744	8,750	20,739
Arunachal Pr	240	198	186	Manipur	263	237	235
Assam	5,783	1,366	5,010	Meghalaya	223	192	178
Bihar	3,332	1,388	2,782	Mizoram	206	206	204
Chandigarh	231	106	177	Nagaland	194	193	191
Chhattisgarh	4,749	1,226	4,403	Odisha	3,286	1,151	2,835
DNH& DD*	2,792	920	1,804	Puducherry	769	371	639
Delhi	3,108	1,206	2,181	Punjab	13,262	2,749	11,375
Goa	704	377	667	Rajasthan	10,614	3,403	9,567
Gujarat	31,157	7,452	24,811	Sikkim	91	90	82
Haryana	10,664	3,464	7,935	Tamil Nadu	39,908	14,827	31,517
Himachal Pr	2,654	916	1,980	Telangana	13,139	3,173	10,700
J & K	1,021	487	838	Tripura	707	699	688
Jharkhand	2,925	867	2,224	Uttar Pr	19,288	6,417	16,263
Karnataka	14,577	5,069	11,391	Uttarakhand	2,890	1,215	2,373
Kerala	7,636	2,410	6,852	West Bengal	10,345	2,834	8,881
Ladakh	33	30	26	All-India	255,244	79,543	206,523

Source:MoSPI (2024a)

1.6.3 *Urban Frame Survey*

For conducting a sample survey of households or establishments, it is not feasible to maintain an up-to-date sampling frame of these ultimate units (i.e., households/establishments) for the purpose of drawing samples in a vast country like India. The alternative is to develop and maintain a frame of area units from where a representative area sample can be drawn to prepare the list of households/establishments in the selected areas during the survey and draw a random sample of households or establishments from this list for detailed enquiry. In the rural areas, list of villages as per the population census serves as the area frame. The primary objective for launching the Urban Frame Survey (UFS) has been to provide an up-to-date sampling frame of compact area units (i.e., urban blocks) for the urban areas with clear identifiable boundaries, which can be considered as the counterparts of villages in the rural areas, to draw a random sample of such blocks in the socio-economic surveys.

Statement 1.4 State-wise total number of UFS blocks

State/UT	Number of UFS Blocks	State/UT	Number of UFS Blocks
Andaman & Nicobar Is	247	Madhya Pradesh	38,627
Andhra Pradesh	29,472	Maharashtra	81,796
Arunachal Pradesh	576	Manipur	1,445
Assam	7,611	Meghalaya	1,102
Bihar	16,895	Mizoram	777
Chandigarh	2,570	Nagaland	952
Chhattisgarh	9,255	Odisha	12,931
DNH& DD*	688	Puducherry	1,308
Delhi	30,002	Punjab	20,331
Goa	1,592	Rajasthan	32,391
Gujarat	44,828	Sikkim	324
Haryana	23,313	Tamil Nadu	65,109
Himachal Pradesh	2,353	Telangana	26,644
Jammu & Kashmir	5,798	Tripura	1,234
Jharkhand	15,382	Uttarakhand	4,582
Karnataka	39,532	Uttar Pradesh	71,524
Kerala	30,859	West Bengal	50,648
Ladakh	105	All-India	672,873
Lakshadweep	70		

*Dadra and Nagar Haveli & Daman and Diu
Source MoSPI (2025b)

The blocks in the UFS are formed by first dividing the town, which is a large geographical entity, into a few compact geographical areas having nearly 20,000 population in each, called 'Investigator Units' (IV Units) with each IV Unit containing about 20 to 40 urban blocks and finally dividing each IV Unit into certain number of urban (UFS) blocks each containing on an average about 120–150 households or roughly 600 to 800 population in general. Statement 1.4 presents the State/UT-wise total number of UFS blocks.[21] As per the latest information available,[22] the total number of UFS blocks in the country is about 750,000.

The UFS blocks are formed with well-defined boundaries (natural boundaries to the extent possible). The IV Unit/UFS block maps depict lanes, by-lanes, railway lines, canals and other landmarks which constitute the boundaries of IV Units/UFS blocks. They also show corner points facilitating easy identification of the blocks during the socio-economic surveys. The UFS blocks so formed are mutually exclusive and exhaustive. While carving out UFS blocks, care is taken to see that it leaves no

[21] MoSPI (2025b).

[22] MoSPI (2025a).

area so that total area of the UFS blocks formed within a town add up to the total area of the town.

Coming to the history of UFS, in the initial rounds of the NSS till the fifteenth round (July 1959–June 1960), the list of census enumeration blocks (CEBs) demarcated during the previous population census was being used in the NSS for selection of urban blocks. However, the use of urban CEBs as the frame led to increased difficulty in the field to identify the boundaries of the sample blocks in many cases mainly due to two reasons: one, the lack of clear boundaries of the CEB maps in certain cases and two, as Murthy and Roy (1975) mentions,[23] the CEBs are 'usually described in terms of census house numbers which often get defaced or erased over a period of time'. Accordingly, the need for a better sampling frame for the urban areas led to the initiation of the UFS in 1959–60.

Initially, the exercise of the UFS was confined to Class I towns / cities only. The first UFS was conducted during 1959–63. Since the sixteenth round of the NSS (July 1960–August 1961), these UFS blocks (along with the census enumeration blocks where UFS was not conducted) were being used as the sampling frame for the selection of urban blocks. With the rapid urbanization it was felt necessary to update the UFS frame on a regular basis. Accordingly, after the NSSO came into existence, the work of UFS was taken up as a regular exercise of the NSSO since 1972 to update the UFS frame in a period of five years by considering all cities and towns of the country. The time periods for the subsequent UFS phases are as follows: 1972–77, 1977–82, 1982–87, 1987–92, 1992–97,1997–2002, 2002–07, 2007–12, 2012–17, 2017–22, and 2022–27 (current phase that is continuing). While the UFS maps were prepared by using pen & paper till 2012–17 phase, UFS Phase 2017–22 adopted a digital mode for the first time using the mobile and web applications. According to MoSPI,[24] 'geo-referenced block/ward/IV unit/town boundaries along with attribute data' are captured in the UFS phase 2017–22. The digital UFS maps are likely to help in better identification of the urban blocks in the surveys.

For the conduct of UFS, the concept of urban area (i.e., town) in the UFS remains exactly the same as adopted in the preceding population census. As per the Census 2011, the urban area includes (a) all the places having a municipality, corporation, cantonment board or notified town area committee, etc. and (b) a place satisfying the following three criteria simultaneously: (i) a minimum population of 5000; (ii) at least 75% of the male working population engaged in non-agricultural pursuits; and (iii) a population density of at least 400 per sq. km. In addition, the outgrowths of cities and towns are also treated as urban as part of Urban Agglomerations.

Availability of some auxiliary information in respect of the units constituting the universe always helps in stratifying the universe and in the sampling of units. To achieve this purpose, the list of UFS blocks, serving as the urban sampling frame of primary sampling units in most of the NSS rounds, provides two important auxiliary information in respect of each UFS Block namely the number of households in the UFS block and 'type of area' to which the UFS block belongs. While the number

[23] Murthy and Roy (1975).

[24] MoSPI (2025e).

of households can be used as the 'size measure' for sampling of UFS blocks if so required (for example, in the case of sampling of UFS blocks by probability proportional to size), the type of area can serve as a useful stratification variable for stratifying the UFS blocks.

As per the existing practice being followed in the UFS considerably over a long time period, the different types of area to which a UFS block can belong include: Affluent Area (introduced since UFS 2007–12), Residential Area other than Affluent Area, Business/Bazaar Area, Slum Area, Industrial Area, Hospital Area, Educational/ Institutional Area, Prohibited Area, and Other Area Not Elsewhere Classified. The UFS has the provision to assign at the most two different types of area to a particular UFS block. In the survey on living standard of the people/households, like the survey on household consumer expenditure, the UFS blocks categorized as 'affluent area' can together form one sub-stratum within a given stratum (like district/town or group of towns) so that part of the sample size of total number of blocks can be assigned to this sub-stratum for ensuring adequate number of rich or relatively affluent households in the overall sample. Similarly, the blocks identified as 'slum area' can together form one stratum/sub-stratum in the survey of slum dwellers or other related surveys where the survey should ensure a fair representation of households residing in the urban slums. In case of the establishment surveys, separate strata/sub-strata of those UFS blocks can be formed which are identified as either a Business/Bazaar Area or Industrial Area or Hospital Area or Educational/Institutional Area or a combination of them depending upon the industries under the survey coverage.

1.6.4 Sample Checks on Area Enumeration and Crop Estimation Surveys

The responsibility for collection of crop statistics and compilation of crop-wise area figures and the yield rates of principal food and non-food crops rests with the respective State/UT Governments. As regards the compilation of area figures, States/UTs can be grouped into three categories namely, (i) land record States/UTs (24 in number)[25] where area figures are generated based on complete enumeration i.e., aggregation of village-level area figures which are noted down by '*Patwaris*' in a register popularly known as 'Khasra Register'; (ii) non-land record States (7 in number)[26] with area statistics generated based on the sample survey of 20% villages in each year; and (iii) the remaining States/UTs/areas where a satisfactory system of collection of area statistics is not yet developed. While the States/UTs under the

[25] These States/UTs include Andhra Pradesh, Assam, (except two hill districts of North Cachar and Karbi Anglong), Bihar, Chandigarh, Chhattisgarh, Daman & Diu, Dadra & Nagar Haveli (erstwhile), Delhi, Goa, Gujarat, Haryana, Himachal Pradesh, Jammu & Kashmir, Jharkhand, Karnataka, Madhya Pradesh, Maharashtra, Puducherry, Punjab, Rajasthan, Tamil Nadu, Telangana, Uttarakhand and Uttar Pradesh (together with Andaman district of Andaman & Nicobar Islands) [MoSPI (2023a)].

[26] Arunachal Pradesh, Kerala, Nagaland, Odisha, Sikkim, Tripura and West Bengal [ibid.].

first category cover about 86% of the all-India reporting area, those under the second category account for nearly 9% of the reporting area.[27]

So far as the statistics on yield rate is concerned, the same is derived by conducting scientifically designed Crop Cutting Experiments (CCE) as part of the General Crop Estimation Surveys (GCES). The sample design generally adopted for the Crop Estimation Survey is stratified multi-stage random sampling with *tehsils/talukas* as strata and village within a stratum as first-stage unit of sampling, survey number/field within each selected village as sampling unit at the second stage and experimental plot of a specified shape and size as the ultimate sampling unit. The total number of CCEs at the country level[28] stood at 17,49,586 during GCES 2019–20.

The NSSO through its Agricultural Statistics Headquarters in the Field Operations Division (FOD) is entrusted with the twin responsibilities: firstly, coordinating the GCES results of State Government and ensuring adherence to statistical standards to maintain comparability and reliability; and secondly, guiding the States to provide reliable estimates of area and yield rates through the scheme known as the Improvement of Crop Statistics (ICS). This scheme was introduced way back in 1973–74 with a special emphasis on wheat in nine major wheat growing States. Subsequently, over time, the coverage was extended by including other major crops and other States. Currently, the scheme is operational in 20 States and 2 UTs[29] covering about 44 crops. The objective of the scheme is to find out—through the joint efforts of the NSSO and the State Governments—deficiencies in crop statistics and suggest remedial measures. For this purpose, sample checks on area enumeration are conducted in a sample of about 10,000 villages in each season and a little more than 31,000 CCEs are conducted in a year (covering 20 States and 2 UTs at present) to check the quality of primary data. Approximately half of this total overall non-overlapping sample is covered by the NSSO and the other half by the State Agricultural Statistics Authorities (SASAs).[30] Based on the findings of the sample checks under this scheme, Agricultural Statistics Headquarters of FOD, NSSO brings out an annual publication titled 'Review of Crop Statistics System in India through the Scheme of Improvement of Crop Statistics' and these findings including the estimates of yield

---

[27] Ibid.

[28] MoSPI (2023b).

[29] These States and UTs are Andhra Pradesh, Assam, Bihar, Chhattisgarh, Gujarat, Haryana, Himachal Pradesh, Jammu & Kashmir, Jharkhand, Karnataka, Kerala, Madhya Pradesh, Maharashtra, Odisha, Puducherry, Punjab, Rajasthan, Tamil Nadu, Telangana, Uttarakhand, Uttar Pradesh and West Bengal.

[30] Distribution of overall sample size of CCE (Extracts from Sect. 6.2 of the MoSPI Publication "Review of crop statistics system in India through the scheme of improvement of crop statistics", 46th issue, 2020–21): "The total number of experiments to be supervised by each agency under the scheme is distributed among the specified crops and States taking into account mainly the crop area and availability of man power resources. District is the 'Stratum' for supervision of crop cutting experiments. Within a State, the experiments are allotted to the strata and crops in proportion to the number of crop cutting experiments planned under GCES. The number of villages in a district for inspecting experiments on a crop is determined at the rate of two experiments per village, except for minor crops in Odisha where it is one experiment per village."

rates based on the sample checks are used by the Ministry of Agriculture & Farmers Welfare to validate the estimates they receive from the State Governments/SASAs.

References

Government of India. (1951). *First report of the national income committee.*

Government of India. (1954). *Final report of the national income committee.*

Government of India. (2001). *Report of the national statistical commission.*

MoSPI. (2008a). *National industrial classification (all economic activities), 2008.* Central Statistical Organisation, Ministry of Statistics & Programme Implementation, Government of India.

MoSPI. (2008b). *Manual on annual survey of industries.* Central Statistical organization, Ministry of Statistics & Programme Implementation, Government of India.

MoSPI. (2009). *Statistical system in India 2009.* Ministry of Statistics & Programme Implementation, Government of India.

MoSPI. (2011). *Annual survey of industries, 2008–09* (Vol. I). Central Statistics Office, Ministry of Statistics & Programme Implementation, Government of India.

MoSPI. (2012). *Annual survey of industries, 2009–10* (Vol. I). Central Statistics Office, Ministry of Statistics & Programme Implementation, Government of India.

MoSPI. (2013). *Annual survey of industries, 2010–11* (Vol. I). Central Statistics Office, Ministry of Statistics & Programme Implementation, Government of India.

MoSPI. (2015). *Changes in methodology and data sources in the new series of national accounts, base year 20ll-12.* Central Statistics Office, Ministry of Statistics & Programme Implementation, Government of India.

MoSPI. (2016a). *Annual survey of industries*, 2013–14 (Vol. I). Central Statistics Office, Ministry of Statistics & Programme Implementation, Government of India.

MoSPI. (2016b). *Annual survey of industries*, 2011–12 (revised) (Vol. I). Central Statistics Office, Ministry of Statistics & Programme Implementation, Government of India.

MoSPI. (2017). *Annual survey of industries*, 2014–15 (Vol. I). Central Statistics Office, Ministry of Statistics & Programme Implementation, Government of India.

MoSPI. (2023a). *Review of crop statistics system in India through the scheme of improvement of crop statistics: 46th issue, 2020–21.* National Sample Survey Office (Field Operations Division), Agricultural Statistics Headquarters, Government of India.

MoSPI. (2023b). *Consolidated results of crop estimation survey on principal crops 2019–20.* National Sample Survey Office (Field Operations Division), Agricultural Statistics Headquarters, Government of India.

MoSPI. (2024a). *Annual survey of industries: 2022–2023, Summary results for factory sector.* National Statistics Office, Enterprise Survey Division, Government of India.

MoSPI. (2024b). *Annual survey of industries: 2022–2023* (Volume-I). National Statistics Office, Enterprise Survey Division, Government of India.

MoSPI. (2025a). *Diamond jubilee publication, household surveys of NSS, a journey of 75 years.* Government of India.

MoSPI. (2025b). *Urban frame survey.* https://mospi.gov.in/ufs-metadata. Accessed from the website on March 17, 2025.

MoSPI. (2025c). *About ministry.* Ministry of Statistics & Programme Implementation, Government of India. Accessible at https://www.mospi.gov.in/about-ministry-0.

MoSPI. (2025d). *Offices of NSSO (FOD) in India.* Ministry of Statistics & Programme Implementation, Government of India. Accessible at https://www.mospi.gov.in/office-finder-nsso.

MoSPI. (2025e). *Urban frame survey. Ministry of Statistics & Programme Implementation, Government of India.* Accessible at https//mospi.gov.in/urban-frame-surveyufs.

Murthy, M. N., & Roy, A. S. (1975). Development of the sample design of the Indian national sample survey during its first twenty-five rounds. *Sankhya*, *37*, (Series C, Pt. 1), 1–42.

Chapter 2
Salient Features of the NSS

2.1 Concept of NSS Round

Every survey carried out as part of the National Sample Survey (NSS) is assigned a round number. With the establishment of the NSS in the year 1950, the fieldwork of the first such survey i.e., the NSS 1st round was conducted during October 1950 to March 1951. The fieldwork for each survey/NSS round spreads over a certain duration. Most of the NSS rounds have been either of one-year duration or of six months' duration. Recently, apart from the NSS rounds devoted to different subjects of enquiry, some surveys like the Periodic Labour Force Survey (PLFS) and the Annual Survey of Unincorporated Sector Enterprises (ASUSE) are being carried out on an annual basis without assignment of any round numbers to such surveys.[1]

2.2 Randomness of the Sample and the Issue of Inaccessibility

Generally, a stratified multi-stage (two or three stage) design has been adopted in the NSS for its socio-economic enquiries with villages and urban blocks comprising the primary (first-stage) sampling units. The NSS takes due care to ensure randomness of the sample. As a result, some of the primary sampling units, particularly the villages, could be located in far-flung areas with difficulty to reach. For example, in the 1st round of the NSS, with an overall sample of 1,833 villages scattered all over the

[1] The readers may please note that the Indian National Sample Survey (NSS) generally uses the term 'enterprises' in its various rounds on the subject including the ASUSE. However, the ultimate sampling units in these surveys are actually the 'establishments' located in single locations and not the 'enterprises' where each enterprise will consist of more than one establishment in case of multi-establishment enterprises having branches in different places. In this book, the terms 'enterprise' and 'establishment' have been used interchangeably to maintain consistency with the NSS terminology.

G C Manna, *75 Years of the Indian National Sample Survey*, India Studies in Business and Economics, https://doi.org/10.1007/978-981-96-7320-9_2

country, some villages were located in areas which were difficult to access. As the NSS Report (1952) mentions,[2] 'For instance, there were villages in the wild areas of Orissa such as Kalahandi where the investigators had to go through more than 20 miles of wild forests under the protection of armed guards. For another sample village in northern Uttar Pradesh an investigator had to wait for the snow to melt to go over the high passes of the Himalayas. Some villages of the tribal areas also came within the sample' (Ministry of Finance, 1952).

2.3 Geographical Coverage

Since NSS 69th round (July 2012–December 2012), the NSS has been covering entire geographical area of the country except some villages in Andaman & Nicobar Islands which are difficult to access. Previously, for quite some years, the NSS did not cover a few other areas also due to some operational issues.[3] For example, the interior villages of Nagaland located beyond 5 km of the bus route apart from the villages in Andaman & Nicobar Islands which remain inaccessible throughout the year were not under the coverage of the NSS for quite a few years/rounds till NSS 68th Round (July 2011–June 2012). Similarly, Ladakh and Kargil districts were also not covered for quite a long period till NSS 64th Round (July 2007–June 2008). Going back to some earlier periods, for example, NSS 27th round (October 1972–September 1973), i.e., the first survey in the series of quinquennial surveys on employment-unemployment and household consumer expenditure, did not cover 'Ladakh district of Jammu & Kashmir, rural areas of Pal and Sumri tehsils of district Surguja and Bjapur, Dantewara, Konta and Narayanpur tesils of district Bastar of Madhya Pradesh, Rajura Sironcha and Godh-Chiroli tehsils of district Chanda and Melghat tehsil of district Amaravati of Maharashtra, rural areas of Nagaland, disturbed villages of Tripura, rural areas of the Union Territory of Chandigarh, the union territories of Arunachal Pradesh, Andaman and Nicobar Islands, Mizoram, Laccadive, Minicoy and Amindivi Islands and Dadra & Nagar Haveli[4]'. By the third survey in this quinquennial series i.e., NSS 38th round (January–December, 1983), the survey coverage improved and this round excluded Ladakh & Kargil districts of Jammu & Kashmir and rural areas of Nagaland.[5]

[2] Refer to Ministry of Finance (1952), Chapter Three, Sect. 5.1. See also Rao (2016).

[3] However, the excluded areas had a very small share in the total population of the country.

[4] See National Sample Survey Organisation's 27th Round (October 1972–September 1973) Report, Number 284, Government of India. Accessible at https://mospi.gov.in/sites/default/files/publication_reports/nss_rep_284_0.pdf.

[5] Department of Statistics (1985).

2.4 Rural-Urban Classification

The NSS adopts exactly the same rural and urban categorization of the geographical areas as adopted in the previous Population Census. As per Census 2011, urban areas comprise two types of administrative units i.e., statutory towns and census towns. Statutory towns include 'all administrative units that have been defined by statute as urban like Municipal Corporation, Municipality, Cantonment Board, Notified Town Area Committee, Town Panchayat, Nagar Palika etc.' while Census towns are composed of administrative units satisfying the following three criteria simultaneously: 'i) It should have a minimum population of 5,000 persons; ii) At least 75 per cent of the male main working population should have been engaged in non-agricultural pursuits; and iii) It should have a density of population of at least 400 persons per sq. km. (1,000 per sq. mile)'. Accordingly, any administrative area not classified as urban is treated as a rural area.

The lowest geographical unit in rural area for the sampling purposes in the NSS is the census village. It is worthwhile to mention that if some rural areas are subsequently (i.e., subsequent to the census) declared as urban by appropriate notifications and for which the Urban Frame Survey (UFS) has been completed, such areas are also treated as urban in the NSS after including them in the UFS frame for sampling purposes.

2.5 Subjects of Enquiry

Over its various 'rounds' the NSS has covered a number of subjects of current interest through its household and enterprise surveys. Some of the important subjects of enquiry include household consumer expenditure, employment-unemployment including migration, land and livestock holdings, debt and investment, housing condition, urban slums, drinking water including sanitation & hygiene, social consumption covering health and education, survey of disabled persons and non-agricultural enterprises covering manufacturing, trading as well as other service sector enterprises.

While deciding the subjects of enquiry, requests of central government ministries and other users are also taken into consideration. For example, as per the request of the Ministry of Agriculture & Farmers Welfare, a 'situation assessment survey of farmers' was conducted as part of NSS 59th round (January–December 2003). The survey was repeated in NSS 70th Round (January 2013–December, 2013) and subsequently in NSS 77th round (January–December 2019) as 'Situation Assessment Survey of Agricultural Households'. Similarly, accommodating the request of Ministry of Tourism, Government of India, a comprehensive survey on domestic tourism was conducted for the first time during NSS 65th round (July 2008–June 2009) and the survey was repeated in NSS 72nd round (July 2014–June 2015) to facilitate the Ministry in preparing Tourism Satellite Account.

Interestingly, some of the subjects were so far covered in one or only a very few rounds. For example, the Survey on the practices of 'jhum' cultivation was conducted in NSS 31st round (July 1976–June 1977), Survey on living condition of tribals in NSS 44th round (July 1988–June 1989) and Use of common property resources in NSS 54th round (January–June 1998). Further, there are some subjects which were not taken up in the NSS after certain point of time like land utilization and yield/crop surveys conducted in some initial rounds but not taken up after 24th round (July 1969–June 1970); Income and expenditure being last covered in NSS 14th round (July 1958–June 1959), Livestock survey in NSS 30th round (July 1975–June 1976) and Population, births and deaths as part of NSS 39th round (January–June 1984).

2.6 Subjects Covered in Different NSS Rounds

NSS conducted its latest i.e., 79th round during July 2022–June 2023. The fieldwork for NSS 80th round is being carried out during January–December 2025. During its 75 years, NSS has covered a large number of subjects for meeting the data requirements of a variety of users including the government organizations. Statement 2.1 describes the subjects covered in different rounds of NSS since its 1st round. It may be seen that recently, apart from the subjects covered though various NSS rounds, quite a few additional surveys have been conducted by the NSSO like the Periodic Labour Force Survey (PLFS), Time Use Survey (TUS) and the Annual survey of Unincorporated Sector Enterprises (ASUSE) to bridge the data gaps on certain subjects.

Statement 2.1 Subjects of enquiry in different NSS rounds

Round No.	Period of survey	Subjects covered
1	October 1950–March 1951	Wages, land utilization, household enterprises, household assets and liabilities and monthly working account for Pune Schedules, consumer expenditure, prices
2	Apr'51–June'51	Village statistics, agriculture, consumer expenditure
3	Aug'51–Nov'51	Village statistics, agriculture and animal husbandry, small-scale manufacture and handicrafts, trade, profession, services and financial operations, consumer expenditure, prices
4	Apr'52–Sep'52	Village statistics, land utilization, agriculture and animal husbandry, small scale manufacture and handicrafts, transport, trade, profession service and financial operations, consumer expenditure, prices

(continued)

(continued)

Round No.	Period of survey	Subjects covered
5	Dec'52–Mar'53	Land utilization, agriculture and animal husbandry, non-household manufacturing establishments, small-scale manufacture and handicrafts, transport, trade, profession service and financial operations, consumer expenditure, prices
6	May'53–Sep'53	Village statistics, land utilization, agriculture and animal husbandry, non-household manufacturing establishments, small-scale manufacture and handicrafts, transport, trade, profession service and financial operations, consumer expenditure, prices, opinion of newspaper readers about newspaper reading
7	Oct'53–Mar'54	Village statistics, land utilization, agriculture and animal husbandry, non-household manufacturing establishments, small scale manufacture and handicrafts, transport, trade, profession service and financial operations, consumer expenditure, prices, opinion on out-turn of rice in village
8	Jul'54–Mar'55	Housing condition, land holdings, trend of self-management of agricultural holdings, land utilization, small-scale manufacture and handicrafts, transport, trade, profession service and financial operations, household indebtedness, consumer expenditure, prices, farming prices
9	May'55–Nov'55	Village statistics, employment and unemployment, land utilization, small scale manufacture and handicrafts, transport, trade, consumer expenditure, income and expenditure, prices
10	Dec'55–May'56	Village statistics, housing condition, employment and unemployment, land utilization and yield survey, small scale manufacture and handicrafts, transport, trade, profession service and financial operations, income and expenditure, prices
11	Aug'56–Jan'57	Village statistics, housing condition, employment and unemployment and indebtedness of agricultural labour households, employment and unemployment of households other than agricultural labour households, land utilization and yield survey, agriculture and animal husbandry, income and expenditure, prices, weights and measures in urban areas

(continued)

(continued)

Round No.	Period of survey	Subjects covered
12	Mar'57–Aug'57	Village statistics, vital statistics, housing condition, employment and unemployment and indebtedness of agricultural labour households, employment and unemployment of households, other than agricultural labour households, land utilization and yield survey, production of milk and production and utilization of cattle dung, housing, income and expenditure, prices, weights and measures in urban areas
13	Sep'57–May'58	Village statistics, vital statistics, housing condition, employment and unemployment, land utilization and yield surveys, income and expenditure, prices, and reader's preference
14	Jul'58–June'59	Village statistics, population, births and deaths, employment, land utilization and yield surveys, small-scale manufacture and handicrafts, income and expenditure, prices
15	Jul'59–June'60	Population, births and deaths, housing condition, employment and unemployment, land utilization and yield surveys, non-mechanized transport and utilization of working animals, livestock products, non-registered expenditure, prices, disposal of cereals by producer households
16	Jul'60–June'61	Population, births and deaths, family planning, housing condition, employment and unemployment, urban labour force, number of physically handicapped persons, ownership of land and operational holdings (in rural only), land utilization and yield surveys, household indebtedness, consumer expenditure, prices
17	Jul'61–June'62	Population, births and deaths, family planning, morbidity, employment, urban labour force, ownership of land and operational holdings, land utilization, yield surveys, capital formation, consumer expenditure, prices
18	Feb'63–Jan'64	Village statistics, population, births and deaths, migration, housing condition, urban labour force, land utilization, yield survey, earnings from professions and liberal arts, construction, indebtedness of scheduled tribe households in Manipur and Tripura, income of rural labour households, consumer expenditure, prices

(continued)

(continued)

Round No.	Period of survey	Subjects covered
19	Jul'64–June'65	Village statistics, population, births and deaths, housing condition, urban labour force, employment, unemployment and indebtedness of rural labour households, land utilization, yield survey, integrated household schedule detailed and abridged, prices
20	Jul'65–June'66	Village and block statistics, population, births and deaths, housing condition, urban labour force, employment and unemployment and indebtedness of rural labour households, land utilization, yield survey, integrated household schedule with emphasis on trade and land utilization, prices
21	Jul'66–June'67	Village and block statistics, population, births and deaths, housing condition, urban labour force, integrated households schedule detailed and abridged with emphasis on land utilization, prices, opinion on production of cereal crops
22	Jul'67–June'68	Village and block statistics, population, births and deaths, housing condition, number of pucca houses, urban labour force, land utilization, integrated household surveys, prices, farming practices, opinion on production of cereal crops
23	Jul'68–June'69	Population, births and deaths, housing condition, number of pucca houses, land utilization, yield survey, small-scale manufacturing (household and non-household), integrated household schedule, prices, opinion on production of cereal crops
24	Jul'69–June'70	Number of pucca houses, number of physically handicapped persons, land utilization, yield survey, non-registered distributive trade, integrated household schedules, prices, opinion on production of cereal crops
25	Jul'70–June'71	Statistical check on land utilization, indebtedness of non-manual employee households in urban areas, economic conditions of weaker section of rural population, integrated household schedule (revised), prices
26	Jul'71–Sep'72	Village statistics, number of pucca houses, land holdings, statistical check on land utilization, debt and investments, consumer expenditure, prices
27	Oct'72–Sep'73	Seasonal migration, number of pucca, semi-pucca and kutcha houses, employment and unemployment, current building activity in rural areas, consumer expenditure, prices

(continued)

(continued)

Round No.	Period of survey	Subjects covered
28	Oct’73–June’74	Population, births and deaths, morbidity, fertility, maternal and child care, family planning, housing condition, normal health facilities in rural areas, disability, number of physically and mentally affected persons, number of new buildings, consumer expenditure, prices
29	Jul’74–June’75	Employment, unemployment and indebtedness of rural labour households, small scale manufacture and handicrafts, mining and quarrying, trade, hotels and restaurants, transport, service, construction, consumer expenditure, income, availability and extent of utilization of electricity in rural areas, prices
30	Jul’75–June’76	Village statistics, livestock number and products, consumption of livestock products, livestock enterprises, prices, survey on railway travel
31	Jul’76–June77	Education in both rural and urban areas, survey on the practice of Jhum cultivation in rural areas, household indebtedness survey in Himachal Pradesh, economic condition
32	Jul’77–June’78	Employment and unemployment, consumer expenditure, integrated household survey in North-Eastern region, rural retail prices
33	Jul’78–June’79	Manufacture of cane and bamboo products for own consumption and household weaving in North East region, unorganized manufacture, rural retail prices
34	Jul’79–June’80	Education, medical and health, unorganized sectors of trade, transport, hotels and restaurants, storage, warehousing and services, prices
35	Jul’80–June’81	Maternal and childcare, family planning, education, medical and health, construction activity and social consumption, prices
36	Jul’81–Dec’81	Survey on disabled persons
37	Jan’82–Dec’82	Land holdings and livestock holdings, debt and investment
38	Jan’83–Dec’83	Employment and unemployment, consumer expenditure
39	Jan’84–June’84	Population, births and deaths
40	Jul’84–June’85	Unorganized manufacture-Non-Directory establishments and own account enterprises

(continued)

(continued)

Round No.	Period of survey	Subjects covered
41	Jul'85–June'86	List of households and trading enterprises, trade-non-directory establishments and own account enterprises
42	Jul'86–June'87	Social consumption, problems of aged and Ex-servicemen, household consumer expenditure (short schedule)
43	Jul'87–June'88	Employment and unemployment, consumer expenditure, travel habits of households
44	Jul'88–June'89	Survey on living condition of tribals, housing condition and current construction activities, household consumer expenditure (short schedule)
45	Jul'89–June'90	Non-directory manufacturing establishments & Own account manufacturing enterprises (OAEs), household consumer expenditure (short schedule)
46	Jul'90–June'91	Non-directory manufacturing establishments & Own account trading enterprises (OAEs), household consumer expenditure (short schedule)
47	Jul'91–Dec'91	Culture & literacy, disabled persons, household consumer expenditure (short schedule)
48	Jan'92–Dec'92	Land holdings, livestock holding and debt and investment, household consumer expenditure (short schedule)
49	Jan'93–June'93	Housing condition and migration with special emphasis on slum dwellers and household consumer expenditure (short schedule)
50	Jul'93–June'94	Employment and Unemployment and consumer expenditure
51	Jul'94–June'95	Unorganized manufacture—Directory Manufacturing Establishments, Non-directory manufacturing establishments & Own account manufacturing enterprises, consumer expenditure (short schedule)
52	Jul'95–June'96	Survey on health care, participation in education, consumer expenditure and employment and unemployment (short schedule)
53	Jan'97–Dec'97	Survey on own account trading enterprises and non-directory trading establishments, consumer expenditure and employment and unemployment (short schedule)

(continued)

(continued)

Round No.	Period of survey	Subjects covered
54	Jan'98–June'98	Use of common property resources, sanitation, hygiene and services, consumer expenditure, employment and unemployment
SS	Aug'98–June'99	Unorganized enterprises
55	Jul'99–June 2000	Consumer expenditure, employment and unemployment, informal sector enterprises
56	Jul 2000–June'01	Unorganized manufacturing enterprises, consumer expenditure (thin sample survey)
57	Jul'01–June'02	Unorganized service sector enterprises (excluding trade and finance), consumer expenditure (thin sample survey)
58	Jul'02–Dec'02	Village facilities, slum, disabled persons, housing condition, consumer expenditure (thin sample survey)
59	Jan'03–Dec'03	Situation assessment survey of farmers, land and livestock holdings, debt and Investment, consumer expenditure (thin sample survey)
60	Jan'04–June'04	Morbidity, health care, condition of the aged, consumer expenditure and employment-unemployment (thin sample survey)
61	Jul'04–June'05	Household consumer expenditure, employment-unemployment
62	Jul'05–June'06	Unorganized manufacturing enterprises, consumer expenditure and employment-unemployment (thin sample survey)
63	Jul'06–June'07	Service sector enterprises (excluding trade), household consumer expenditure (thin sample survey)
64	Jul'07–June'08	Employment & unemployment and migration particulars, education, household consumer expenditure (thin sample survey)
65	Jul'08–June'09	Domestic tourism, housing condition, slum
66	Jul'09–June'10	Household consumer expenditure, employment and unemployment
67	Jul'10–June'11	Unincorporated non-agricultural enterprises belonging to manufacturing, trade and other services (excluding construction)
68	Jul'11–June'12	Household consumer expenditure, employment and unemployment
69	Jul'12–Dec'12	Drinking water, sanitation and hygiene, housing condition, urban slums

(continued)

(continued)

Round No.	Period of survey	Subjects covered
70	Jan'13–Dec'13	Situation assessment survey of agricultural households, land and livestock holdings, debt and investment
71	Jan'14–June'14	Health, education
72	Jul'14–June'15	Household consumer expenditure, domestic tourism, household expenditure on services and durable goods
73	Jul'15–June'16	Unincorporated non-agricultural enterprises belonging to manufacturing, trade and other services (excluding construction)
74	Jul'16–June'17	Services sector enterprises (List frame-based survey of 'bigger' enterprises)
75	Jul'17–June'18	Household consumer expenditure, health, education
PLFS	Jul'17–June'18	Employment-unemployment
76	Jul'18–Dec'18	Drinking water, sanitation, hygiene and housing condition, survey of persons with disabilities
PLFS	Jul'18–June'19	Employment-unemployment
77	Jan'19–Dec'19	Land and livestock holdings of households and situation assessment of agricultural households (in rural areas only), debt and investment
TUS	Jan'19–Dec'19	Time Use Survey (TUS)
PLFS	Jul'19–June'20	Employment-unemployment
78	Jan'20–Aug'21	Multiple indicator survey
PLFS	Jul'20–June'21	Employment-unemployment
ASUSE	Apr'21–Mar'22	Annual Survey of Unincorporated Sector Enterprises
PLFS	Jul'21–June'22	Employment-unemployment
PLFS	Jul'22–June'23	Employment-unemployment
79	Jul'22–June'23	Comprehensive annual modular survey (for indicators related to education, out-of-pocket medical expenditure, use of mobile and internet, financial inclusion, ICT skills, possession of assets, drinking water, sanitation, energy use, birth registration, access to transport facilities etc.) including survey on AYUSH
HCES	Aug'22–Jul'23	Household consumer expenditure
ASUSE	Oct'22–Sep'23	Annual Survey of Unincorporated Sector Enterprises
PLFS	Jul'23–June'24	Employment-unemployment
HCES	Aug'23–Jul'24	Household consumer expenditure

(continued)

(continued)

Round No.	Period of survey	Subjects covered
ASUSE	Oct'23–Sep'24	Annual Survey of Unincorporated Sector Enterprises
TUS	Jan'24–Dec'24	Time Use Survey
PLFS	Jan'25–Dec'25	Employment-unemployment
ASUSE	Jan'25–Dec'25	Annual Survey of Unincorporated Sector Enterprises
80	Jan'25–Dec'25	Household social consumption—health, Comprehensive Modular Survey (CMS)—Telecom & Education#

Notes
Planned survey period for the CMS—Telecom: January to March, 2025 & Education: April to June, 2025
HCES: Household Consumption Expenditure Survey
PLFS: Periodic Labour Force Survey; The earlier series prior to PLFS: 2025 continued till Dec'24
SS: Special Survey
Source Information up to NSS 55th round accessed from MoSPI (2001)

2.7 A Multi-subject Enquiry

Each NSS round normally accommodates more than one subject of interest. However, care is taken not to include so many subjects of enquiry in a given round which is likely to make the sample design a complex one. Devoting each NSS round to a single-subject has hardly any scope to accommodate new subjects of current interest and the other requests from the user ministries.

2.8 Periodicity of Subjects

While for many subjects of enquiry there is no specific periodicity and they are covered in a particular round as per the requirements, some subjects of enquiry in the recent past were taken up with more or less a definite periodicity (see Box 2.1) like once in ten years, or once in five years, or even every year.

Box 2.1 Periodicity of certain subjects of enquiry

Subject	Approximate periodicity	NSS Rounds (Years: July to June unless mentioned otherwise)
Land & livestock holding@ and Debt & investment	Once in 10 years	26 (1971–72); 37 (1982); 48 (1992); 59 (2003); 70 (2013); 77 (2019)
Enterprise Surveys: Unorganized Manufacture	Once in 5 years	33 (1978–79); 40 (1984–85); 45 (1989–90); 51 (1994–95); 55 (1999–2000); 56 (2000–2001); 62 (2005–2006)
Enterprise Surveys: Trading Enterprises	Once in 5 years	34 (1979–80); 41 (1985–86); 46 (1990–91); 53 (1997); 55 (1999–2000)
Enterprise Surveys: Unincorporated Non-agricultural Enterprises	Once in 5 years	67 (2010–11) & 73 (2015–16)^
Household consumer expenditure and Employment-Unemployment ('Large sample')*	Once in 5 years	27 (1972–73); 32 (1977–78); 38 (1983); 43 (1987–88); 50 (1993–94); 55 (1999–2000); 61 (2004–2005); 66 (2009–10); 68 (2011–12)
Household consumer expenditure and Employment-Unemployment ('Thin sample')**	Every year (during the stated period and then discontinued)	42 (1986–87) to 64 (2007–2008)
Employment- Unemployment (through the Periodic Labour Force Survey i.e., PLFS)	Every year (with provision to release key quarterly estimates for urban areas); PLFS: 2025 has revamped the survey design to release key quarterly estimates for rural areas also in addition to the generation of key monthly estimates for both rural and urban areas for the country as a whole.	2017–2018 onwards

@ Survey on land & livestock holding was also conducted during NSS 17th round (1961–62)

^ These are the latest two rounds, with a gap of 5 years in-between, on the survey of unincorporated non-agricultural enterprises covering manufacture, trade and other services. Subsequently, MoSPI has introduced the Annual Survey of Unincorporated Sector Enterprises (ASUSE) covering these subjects on an annual basis since 2021–22

*Popularly known as 'quinquennial surveys'; Survey on household consumer expenditure is also carried out during 2022–23 and 2023–24 through the HCES (Household Consumption Expenditure Survey)

**To meet the data gaps on household consumer expenditure during the intervening periods of large sample rounds, annual surveys on the subject through short questions with a thin sample of 2 to 4 households per village/block was conducted during NSS 42nd round till the 64th round of NSS. In this annual series, key information on employment-unemployment was also collected since NSS 45th round (July 1989–June 1990)

2.9 Sampling Frame of First-Stage Units

For its household surveys and enterprise surveys, the NSS generally adopts a stratified two-stage/three-stage design with villages/urban blocks as the primary sampling units/first-stage units (FSUs)[6] and households/enterprises as the ultimate-stage units (USUs).[7] Sampling frame is a list of sampling units with their complete identification particulars so that the units selected from the frame can be clearly identified at the time of survey. Sampling frame should be up-to-date and complete without any omission of units of the population or duplication of units.

For most of the surveys, particularly in the household surveys, the NSS uses the list of villages as per the latest population census as the sampling frame of FSUs in respect of rural areas. For urban areas, the list of Census Enumeration Blocks (CEBs) as per the Population Census (till the fifteenth round, 1959–60) and thereafter the latest available list of UFS blocks have served as the sampling frame for selection of urban blocks. In the case of enterprises surveys, the list of villages/urban blocks giving information on number of enterprises/workers by industry as per the Economic Census (EC) has been mainly used as the sampling frame for selection of villages/urban blocks. In some of the NSS rounds on enterprise surveys, for the urban areas, this EC database has been used as the sampling frame only for big cities while the latest list of UFS blocks has been used as the sampling frame for other cities and towns.[8] For example, in NSS 67th round (2010–11) survey on unincorporated non-agricultural enterprises, in the urban areas, the database as per EC, 2005 served as

[6] Instead of villages/urban blocks comprising the FSUs, there have been some departures in only a few NSS rounds. For example, during NSS rounds 4 to 7, tehsils were the FSUs in rural areas and villages formed the second-stage units. In the urban areas, during NSS rounds 3 to 8, towns formed the FSUs while urban blocks comprised the second-stage units. From NSS 76th round (July–December 2018), in many rounds devoted to household surveys (PLFS being the exception), large villages with population of 1,000 or more as per Census 2011 are notionally divided into a certain number of sub-units (SU) of more or less equal population during the preparation of frame so that present population of each such SU is not likely to exceed 1,200. For this purpose, Census 2011 population of villages is projected by applying suitable growth rates and the number of SUs formed in a village is determined in advance. SUs are formed in large UFS blocks also in the urban sector with the procedure being similar to that adopted in rural areas except that SUs are formed on the basis of number of households in the UFS blocks since UFS frame does not contain information on population of the UFS blocks. Each UFS block with number of households more than or equal to 200 is divided into a number of SUs. In the remaining i.e., smaller villages/UFS blocks, no SU is formed. In other words, for large villages/UFS blocks, SU becomes the FSU.

[7] In case the sample FSU is small with present population less than 1,200, households/enterprises become the second-stage units (with the design becoming a two-stage one) else the FSU is segmented into a few smaller parts called 'hamlet-groups' for rural areas or 'sub-blocks' for urban areas. A sample of segments is selected randomly to prepare the frame of households/enterprises. In other words, for large FSUs, the design becomes a three-stage design with villages/urban blocks becoming the FSUs while segments and households (or enterprises, as the case may be) forming the second-stage units and third-stage/ultimate stage units respectively.

[8] This was in view of the apprehension that it would be difficult to identify the boundaries of the urban blocks in smaller cities and towns with the help of the maps since a considerable time passed after EC.

the sampling frame only for 26 cities with population more than a million[9] as per census 2001.

Ideally, the sampling frame should contain some auxiliary information about each constituent sampling unit so that this information can be utilized for stratification and sampling of units. The list of villages as per population census contains quite a few such auxiliary information. For example, the village-level database as per census 2011 include information on population (with break-up for SC & ST), number of households, number of literates/illiterates, number of workers (main & marginal) by categories like cultivators, agricultural labourers, workers in household industries, and other workers. Among these variables, the information on population has been used most frequently. The total population of the village has been used as the 'size measure' for sampling of villages either by (a) probability proportional to size (PPS) with replacement in many recent NSS rounds starting with NSS 58th round (July–December 2002) or (b) circular systematic sampling with PPS in quite a large number of rounds till NSS 55th round (July 1999–June 2000).

2.10 NSS Regions

NSS Regions are hierarchical geographical domains below the level of State/UT. Each NSS region comprises a group of districts within a State/UT except a few smaller States/UTs where all the districts of the State/UT together constitute one region. During the first three rounds of NSS, no region was formed. Subsequently, in quite a few NSS rounds, there was no concept of any 'fixed region'. For example: 'From 4 to 10th and 13th to 15th rounds of NSS, 52 natural divisions of 1951 population census which were groups of districts 'having similar geographical features and population densities' were treated as regions...... During the 11th and 12th rounds 39 regions were formed for Agricultural Labour Enquiry (ALE) by combining 75 ALE zones formed by Labour Bureau for conducting the first ALE in 1950–51'.[10]

The need for having a fixed set of regions for the purpose of geographical stratification and also for having a comparable time series estimates for the domains below the State and above the district[11] led to formation of 64 NSS regions[12] in 1965 'in consultation with different Central Ministries, Planning Commission, Registrar General and State Statistical Bureaus, by grouping contiguous districts having similar geographical features, rural population densities and crop-pattern.....These regions were in use up to the 31st round [1976–77]. This set of regions was revised during

[9] These 26 cities excluded Mumbai because of some identification problem of blocks in the city as per the EC.

[10] MoSPI (2001).

[11] District level estimates for many socio-economic indicators as per the overall sample size adopted in the NSS are not likely to be reliable due to limitations of sample size.

[12] Formation of stratum at the NSS region level for the purpose of sampling of first-stage units (i.e., villages/urban blocks) ensures a much more geographical spread of the sample villages/blocks within the State as compared to the situation when strata are formed at the State level.

1977 and total number of regions were increased to 73 in consideration of the changed conditions. This revised set of regions was in use during 32nd [round] and 35th round [1980–81]. The total number of regions went up to 77 during 36th to 43rd rounds [1987–88] after the State/Union Territories of Sikkim, Andaman & Nicobar Islands, Dadra and Nagar Haveli and Lakshadweep were covered in NSS from its 36th round. From NSS 44th round, total number of regions became 78 after Goa was declared a separate state'.[13] The total number of NSS regions increased further over time due to subsequent formation of some more States and districts as well as other operational considerations. As of now, there are 87 NSS regions in the country. Statement 2.2 gives the composition of different NSS regions with the names of the constituent districts.

2.11 Stratification of First-Stage Units

The primary objective of stratification is to increase the efficiency of the survey design by including homogeneous units in relation to the study variable under a given stratum. Stratification with appropriate division of the State into a few strata also ensures much more geographical spread of the sample FSUs within a State. The maximum number of strata that can be formed is dependent upon the overall sample size of FSUs. As compared with the initial NSS rounds, the sample size and also the total number of strata have increased considerably over time in the NSS.

In the initial few rounds, so far as rural area is concerned, strata were formed by grouping contiguous districts. The total number of rural strata at the country level was around 250 between NSS 1st round and 10th round. In the urban area, only a few strata were formed by considering the population of the towns. For example, in NSS 3rd round to 5th round (with the 1st round and 2nd round covering rural areas only), each of the four metro cities formed a separate stratum and all other cities and towns of the country were grouped into three strata only with a total number of 7 urban strata for the country as a whole. In the subsequent three NSS rounds, apart from each of the four metro cities forming a separate stratum, sixteen more strata were formed by forming four strata of remaining cities and towns as per their population size within each of the four census zones and the same resulted in 20 urban strata for the country as a whole. The total number of urban strata increased to 94 in the 9th round.

Generally, each district has been considered as a separate stratum in the rural areas since NSS 28th round (October 1973–June 1974). However, in the urban areas, for a considerable period of time, strata were formed by grouping cities/towns into a few population size classes either within the country or within the census zones or within a State. Subsequently, from NSS 42nd round (1986–87) till 60th round (January–June 2004), urban strata were formed within each NSS region. Stratum

[13] MoSPI (2001).

Statement 2.2 List of NSS regions with their composition*

State/UT	NSS region		Composition (Names of districts)
	Sl. No.	Description	
Andaman & Nicobar Islands	1	Andaman & Nicobar Islands	All districts
Andhra Pradesh	2	Coastal Northern	Srikakulam, Vizianagaram, Visakhapatnam, East Godavari, West Godavari
	3	Coastal Southern	Krishna, Guntur, Prakasam, Sri Potti Sriramulu Nellore
	4	Inland Southern	Y.S.R.(Cuddapah), Kurnool, Anantapur, Chittoor
Arunachal Pradesh	5	Arunachal Pradesh	All districts
Assam	6	Plains Eastern	Lakhimpur, Dhemaji, Tinsukia, Dibrugarh, Sivasagar, Jorhat, Golaghat, Charaideo, Majuli
	7	Plains Western	Kokrajhar, Dhubri, Goalpara, Barpeta, Bongaigaon, Chirang, Kamrup, Kamrup Metropolitan, Nalbari, Baksa, South Salmara Mankachar
	8	Cachar Plain	Karbi Anglong, Dima Hasao, Cachar, Karimganj, Hailakandi, West karbi Anglong
	9	Central Brahamputra Plains	Morigaon, Nagaon, Sonitpur, Darrang, Udalguri, Hojai, Biswanath
Bihar	10	Northern	Pashchim Champaran, Purba Champaran, Sheohar, Sitamarhi, Madhubani, Supaul, Araria, Kishanganj, Purnia, Katihar, Madhepura, Saharsa, Darbhanga, Muzaffarpur, Gopalganj, Siwan, Saran, Vaishali, Samastipur, Begusarai, Khagaria
	11	Central	Bhagalpur, Banka, Munger, Lakhisarai, Sheikhpura, Nalanda, Patna, Bhojpur, Buxar, Kaimur (Bhabua), Rohtas, Aurangabad, Gaya, Nawada, Jamui, Jehanabad, Arwal
Chandigarh	12	Chandigarh	Chandigarh
Chhattisgarh	13	Northern Chhattisgarh	Koriya, Surguja, Surajpur, Balrampur
	14	Mahanadi Basin	Jashpur, Raigarh, Korba, Janjgir-Champa, Bilaspur, Kabeerdham, Rajnandgaon, Durg, Raipur, Mahasamund, Dhamtari, Balodabazar, Gariyaband, Bemetara, Balod, Mungeli

(continued)

Statement 2.2 (continued)

State/UT	NSS region		Composition (Names of districts)
	Sl. No.	Description	
	15	Southern Chhattisgarh	Uttar Bastar Kanker, Bastar, Narayanpur, Dakshin Bastar Dantewada, Bijapur, Kondagaon, Sukama
Dadra & Nagar Haveli and Daman and Diu	16	Dadra & Nagar Haveli and Daman and Diu	All districts
Delhi	17	Delhi	All districts
Goa	18	Goa	All districts
Gujarat	19	South-Eastern	Panch Mahals, Dohad, Vadodara, Narmada, Bharuch, The Dangs, Navsari, Valsad, Surat, Tapi, Chhota Udepur, Mahisagar
	20	Plains Northern	Mahesana, Sabar Kantha, Gandhinagar, Ahmadabad, Anand, Kheda, Arvalli
	21	Dry Areas	Banas Kantha, Patan
	22	Kachchh	Kachchh
	23	Saurashtra	Surendranagar, Rajkot, Jamnagar, Porbandar, Junagadh, Amreli, Bhavnagar, Botad, Dev Bhumi-Dwarka, Gir Somnath, Morbi
Haryana	24	Eastern	Panchkula, Ambala, Yamunanagar, Kurukshetra, Kaithal, Karnal, Panipat, Sonipat, Rohtak, Jhajjar, Gurgaon, Mewat, Faridabad, Palwal
	25	Western	Jind, Fatehabad, Sirsa, Hisar, Bhiwani, Mahendragarh, Rewari
Himachal Pradesh	26	Central	Kangra, Kullu, Mandi, Hamirpur, Una
	27	Trans Himalayan & Southern	Chamba, Lahul & Spiti, Bilaspur, Solan, Sirmaur, Shimla, Kinnaur
Jammu & Kashmir	28	Mountainous	Kathua, Jammu, Samba
	29	Other Hills	Punch, Rajouri, Doda, Ramban, Kishtwar, Udhampur, Reasi
	30	Jhelam Valley	Kupwara, Badgam, Baramula, Bandipore, Srinagar, Ganderbal, Pulwama, Shupiyan, Anantnag, Kulgam
Jharkhand	31	Ranchi Plateau	Garhwa, Lohardaga, Purbi Singhbhum, Palamu, Latehar, Ranchi, Khunti, Gumla, Simdega, Pashchimi Singhbhum, Saraikela-Kharsawan

(continued)

Statement 2.2 (continued)

State/UT	NSS region		Composition (Names of districts)
	Sl. No.	Description	
	32	Hazaribagh Plateau	Chatra, Kodarma, Giridih, Deoghar, Godda, Sahibganj, Pakur, Dhanbad, Bokaro, Hazaribagh, Ramgarh, Dumka, Jamtara
Karnataka	33	Coastal & Ghats	Uttara Kannada, Udupi, Dakshina Kannada
	34	Inland Eastern	Shimoga, Chikmagalur, Hassan, Kodagu
	35	Inland Southern	Tumkur, Bangalore, Mandya, Mysore, Chamarajanagar, Kolar, Chikkaballapura, Bangalore Rural, Ramanagara
	36	Inland Northern	Belgaum, Bagalkot, Bijapur, Bidar, Raichur, Koppal, Gadag, Dharwad, Haveri, Bellary, Chitradurga, Davanagere, Gulbarga, Yadgir
Kerala	37	Northern	Kasaragod, Kannur, Wayanad, Kozhikode, Malappuram, Palakkad
	38	Southern	Thrissur, Ernakulam, Idukki, Kottayam, Alappuzha, Pathanamthitta, Kollam, Thiruvananthapuram
Ladakh	39	Ladakh	Both Leh and Kargil districts
Lakshadweep	40	Lakshadweep	Lakshadweep
Madhya Pradesh	41	Vindhya	Tikamgarh, Chhatarpur, Panna, Satna, Rewa, Umaria, Shahdol, Anuppur, Sidhi, Singrauli
	42	Central	Sagar, Damoh, Vidisha, Bhopal, Sehore, Raisen
	43	Malwa	Neemuch, Mandsaur, Ratlam, Ujjain, Shajapur, Dewas, Dhar, Indore, Rajgarh, Jhabua, Alirajpur
	44	South	Katni, Jabalpur, Narsimhapur, Dindori, Mandla, Chhindwara, Seoni, Balaghat
	45	South Western	Khargone (WestNimar), Barwani, Betul, Harda, Hoshangabad, Khandwa (EastNimar), Burhanpur
	46	Northern	Sheopur, Morena, Bhind, Gwalior, Datia, Shivpuri, Guna, Ashoknagar
Maharashtra	47	Coastal	Thane, Mumbai Suburban, Mumbai, Raigarh, Ratnagiri, Sindhudurg
	48	Inland Western	Pune, Ahmadnagar, Solapur, Satara, Kolhapur, Sangli

(continued)

Statement 2.2 (continued)

State/UT	NSS region		Composition (Names of districts)
	Sl. No.	Description	
	49	Inland Northern	Nandurbar, Dhule, Jalgaon, Nashik
	50	Inland Central	Nanded, Hingoli, Parbhani, Jalna, Aurangabad, Bid, Latur, Osmanabad
	51	Inland Eastern	Buldana, Akola, Washim, Amravati, Wardha, Nagpur, Yavatmal
	52	Eastern	Bhandara, Gondiya, Gadchiroli, Chandrapur
Manipur	53	Plains	Bishnupur, Thoubal, Imphal West, Imphal East
	54	Hills	Senapati, Tamenglong, Churachandpur, Ukhrul, Chandel
Meghalaya	55	Meghalaya	All districts
Mizoram	56	Mizoram	All districts
Nagaland	57	Nagaland	All districts
Odisha	58	Coastal	Baleshwar, Bhadrak, Kendrapara, Jagatsinghapur, Cuttack, Jajapur, Nayagarh, Khordha, Puri
	59	Southern	Ganjam, Gajapati, Kandhamal, Baudh, Subarnapur, Balangir, Nuapada, Kalahandi, Rayagada, Nabarangapur, Koraput, Malkangiri
	60	Northern	Bargarh, Jharsuguda, Sambalpur, Debagarh, Sundargarh, Kendujhar, Mayurbhanj, Dhenkanal, Anugul
Puducherry	61	Puducherry	All districts
Punjab	62	Northern	Gurdaspur, Kapurthala, Jalandhar, Hoshiarpur, Shahid Bhagat Singh Nagar, Amritsar, Tarn Taran, Rupnagar, Sahibzada Ajit Singh Nagar, Pathankot
	63	Southern	Fatehgarh Sahib, Ludhiana, Moga, Firozpur, Muktsar, Faridkot, Bhatinda, Mansa, Patiala, Sangrur, Barnala, Fazilka
Rajasthan	64	Western	Bikaner, Jodhpur, Jaisalmer, Barmer, Jalor, Sirohi, Pali
	65	North-Eastern	Alwar, Bharatpur, Dhaulpur, Karauli, Sawai Madhopur, Dausa, Jaipur, Ajmer, Tonk, Bhilwara
	66	Southern	Rajsamand, Dungarpur, Banswara, Udaipur

(continued)

Statement 2.2 (continued)

State/UT	NSS region		Composition (Names of districts)
	Sl. No.	Description	
	67	South-Eastern	Bundi, Chittaurgarh, Kota, Baran, Jhalawar, Pratapgarh
	68	Northern	Sri Ganganagar, Hanumangarh, Churu, Jhunjhunun, Sikar, Nagaur
Sikkim	69	Sikkim	All districts
Tamil Nadu	70	Coastal Northern	Thiruvallur, Chennai, Kancheepuram, Vellore, Tiruvannamalai, Viluppuram, Cuddalore
	71	Coastal	Karur, Tiruchirappalli, Perambalur, Ariyalur, Nagapattinam, Thiruvarur, Thanjavur, Pudukkottai
	72	Southern	Dindigul, Sivaganga, Madurai, Theni, Virudhunagar, Ramanathapuram, Thoothukkudi, Tirunelveli, Kanniyakumari
	73	Inland	Salem, Namakkal, Erode, The Nilgiris, Dharmapuri, Krishnagiri, Coimbatore, Tiruppur
Telangana	74	Inland North Western	Adilabad, Komaram Bheem, Mancherial, Nirmal, Nizamabad, Kamareddy, Sangareddy, Medak, Siddipet, Medchal-Malkajgiri, Hyderabad, Rangareddy, Vikarabad, Mahbubnagar, Jogulamba, Wanaparthy, Nagarkurnool, Narayanpet
	75	Inland North-Eastern	Jagtial, Peddapalli, Jayashankar, Bhadradri, Mahabubabad, Warangal Rural, Warangal Urban, Karimnagar, Rajanna, Jangaon, Yadadri, Nalgonda, Suryapet, Khammam, Mulugu
Tripura	76	Tripura	All districts
Uttarakhand	77	Uttarakhand	All districts
Uttar Pradesh	78	Northern Upper Ganga Plains	Saharanpur, Muzaffarnagar, Bijnor, Moradabad, Rampur
	79	Central	Jyotiba Phule Nagar, Meerut, Baghpat, Ghaziabad, Gautam Buddha Nagar, Sitapur, Hardoi, Unnao, Lucknow, Rae Bareli, Kanpur Dehat, Kanpur Nagar, Fatehpur, Bara Banki

(continued)

Statement 2.2 (continued)

State/UT	NSS region		Composition (Names of districts)
	Sl. No.	Description	
	80	Eastern	Pratapgarh, Kaushambi, Allahabad, Faizabad, Ambedkar Nagar, Sultanpur, Bahraich, Shrawasti, Balrampur, Gonda, Siddharthnagar, Basti, Sant Kabir Nagar, Maharajganj, Gorakhpur, Kushinagar, Deoria, Azamgarh, Mau, Ballia, Jaunpur, Ghazipur, Chandauli, Varanasi, Sant Ravidas Nagar (Bhadohi), Mirzapur, Sonbhadra
	81	Southern	Jalaun, Jhansi, Lalitpur, Hamirpur, Mahoba, Banda, Chitrakoot
	82	Southern Upper Ganga Plains	Bulandshahr, Aligarh, Mahamaya Nagar, Mathura, Agra, Firozabad, Mainpuri, Budaun, Bareilly, Pilibhit, Shahjahanpur, Kheri, Farrukhabad, Kannauj, Etawah, Auraiya, Etah, Kanshiram Nagar
West Bengal	83	Himalayan	Darjiling, Jalpaiguri, Koch Bihar, Alipurduar, Kalimpong
	84	Eastern Plains	Uttar Dinajpur, Dakshin Dinajpur, Maldah, Murshidabad, Birbhum, Nadia
	85	Southern Plains	North Twenty Four Parganas, Kolkata, South Twenty Four Parganas
	86	Central Plains	Purba Barddhaman, Hugli, Haora, Paschim Barddhaman
	87	Western Plains	Bankura, Puruliya, Paschim Medinipur, Purba Medinipur, Jhargram

Source MoSPI (2001, 2022)

numbers and their composition within each NSS region in NSS 60th round were as follows:

- Stratum 1: All towns with population (as per Census 2001) less than 50,000;
- Stratum 2: All towns with population 50,000 or more but less than 0.2 million;
- Stratum 3: All towns with population 0.2 million or more but less than 1 million;
- Stratum 4, 5, 6, …: Each city/town with population 1 million or more

As a result of the above change in the method of stratification, the total number of strata in both rural and urban areas increased over time. For example, altogether 615 rural strata and 217 urban strata were formed during NSS 54th round (January–June 1998).[14]

[14] Refer to MoSPI (2001).

To facilitate generation of district-wise estimates even for the urban areas, each district (urban part) is also being considered as a separate stratum from NSS 61st round (2004–2005) which covered the subjects of employment-unemployment and household consumer expenditure. However, each city having a population of 1 million or above (threshold being 1.5 million in some rounds/surveys) is treated as a separate stratum and all other cities and towns of the district are being grouped together to form another stratum within the district. Apart from each district being treated as a separate stratum (both rural and urban), further deep stratification has been resorted to in many rounds since NSS 61st round by arranging the villages/towns of a stratum in ascending order of their population and forming n/2 ultimate strata (i.e., sub-strata)—subsequently changed to n/4 ultimate strata/sub-strata in some of the more recent rounds[15]—of roughly equal population (or equal number of households as per UFS in case of urban), where 'n' denotes the sample size of FSUs for a given stratum, with an allocation of 2 sample FSUs (or 4 sample FSUs as the case may be with some of the more recent rounds) to each ultimate stratum. In view of this deep stratification, total number of rural as well as urban strata in the country increased further from NSS 61st round as compared with the situation earlier.

In some rounds/surveys, there are certain exceptions/modifications to the general principle of stratification as described above. For example, the annual Periodic Labour Force Survey introduced since 2017–2018, covering the subject of employment-unemployment, considers NSS region (instead of district) as the basic stratum for both rural and urban areas. However, in the urban areas, each NSS region is further divided into a few strata by grouping cities/towns according to their population as per Census 2011 as follows: stratum 1 comprising all towns with population less than 50,000; stratum 2: all towns with population 50,000 or more but less than 0.3 million; stratum 3: all towns/cities with population 0.3 million or more but less than 1.5 million; and stratum 4, 5, 6,...: each city with a population of 1.5 million or above. In the rural areas, each NSS region/stratum is divided into n/8 sub-strata of roughly equal population ('n' being the number of sample villages allocated to the NSS region) by arranging the villages in ascending order of population. No sub-stratification is done for urban areas in the PLFS.

Similarly, in the enterprise surveys, FSU (village/urban block) level information on number of enterprises and workers by industry as per the previous economic census (EC) is generally used to stratify the FSUs with a view to capturing in the sample adequate number of villages/urban blocks having sufficient of enterprises, particularly those employing hired workers which are popularly described in the NSS as 'establishments', belonging to different industries. To illustrate, NSS 67th round (2010–2011) survey of unincorporated non-agricultural enterprises covering manufacturing, trade and other services (excluding construction) adopted the following stratification[16]:

---

[15] For example, NSS 68th round (2011–2012) covering employment-unemployment and household consumer expenditure.

[16] This section contains information taken from MoSPI (2012). Refer to the report for further details.

> Each district was a basic stratum in both rural and urban areas. However, in case of urban, each city with population of 1 million or more as per Census 2001 forms a separate stratum and all other cities/towns of a district are grouped to form another stratum….Sub-stratification: (i) Rural: There were three sub-strata in the rural sector, viz.: (1) Villages with at least 5 establishments[17] (NDE/DE) [i.e., Non-Directory Establishments with total number of workers including hired workers being 5 or less/Directory Establishments with 6 or more workers including at least one hired worker] under coverage in the manufacturing sector as per EC-2005 information; (2) Remaining villages having at least 5 NDE/DE under coverage in the services sector including trade as per EC-2005 information; (3) Remaining villages of the stratum. For the State(s) where EC-2005 information could not be used as auxiliary information for stratification/sub-stratification due to limitations of EC-2005 frame, each district was sub stratified into 'r/4' sub-strata with a sample allocation of 4 per sub-stratum where 'r' is the sample allocation for the district/stratum. The sub-strata were formed by arranging the villages in terms of population so that total population of each sub-stratum is approximately the same. (ii) Urban, Million plus cities (excluding Mumbai): For each stratum/million plus city, 20 sub-strata were formed as follows: Sub-stratum 1: Blocks with one or more establishment in insurance & pension funding; Sub-stratum 2: Remaining blocks with one or more establishment in storage & warehousing; Sub-stratum 3: Remaining blocks with one or more establishment in accommodation; Sub-strata 4–8: Remaining blocks with one or more establishment in broad activities of manufacturing ….; Sub-strata 9–12: Remaining blocks with one or more establishment in broad activities of trade ….; Sub-strata 13–19: Remaining blocks with one or more establishment in broad activities of other services …excluding the activities covered under sub-strata 1–3; Sub-strata [Sub-stratum] 20: All remaining blocks of the stratum. (iii) Urban, Other cities and towns (including Mumbai): Two sub-strata were formed: Sub-stratum 1: UFS block types: Bazaar area (BA) / Industrial area (IA) / Hospital area / (HA) / Slum area (SA) which are likely to contain relatively higher number of enterprises; Sub-stratum 2: Remaining UFS blocks of the stratum. (MoSPI, 2012)

NSS 70th round (2013)—devoted to situation assessment survey of agricultural households, land & livestock holdings and debt & investment—is another survey where some modifications were made in the stratification. As per the procedure followed since NSS 61st round, this round also treated each district as a stratum separately for rural as well as urban areas. Further, as usual, each million plus city formed a separate stratum while all the remaining towns of the district formed another stratum. However, before forming district level strata (rural), a special stratum was formed in 20 States/UTs[18] consisting of all the villages of the State/UT with population less than 50 as per census 2001. In Andaman & Nicobar Islands included under the said 20 States/UTs, another special stratum was formed comprising all interior and inaccessible villages (with the other special stratum excluding these inaccessible villages to ensure the formation of non-overlapping strata). Similarly, in the rural areas of Nagaland, one special stratum was formed within the State consisting of all the interior and inaccessible villages.

---

[17] In the NSS, the term 'establishment' is generally used to describe the units working with the help of at least one hired worker on a 'fairly regular basis'. The other type of units without any hired worker are generally termed in the NSS as 'own-account enterprises'. See MoSPI (2001).

[18] These included Andaman & Nicobar Islands, Andhra Pradesh, Assam, Bihar, Chhattisgarh, Delhi, Goa, Gujarat, Haryana, Jharkhand, Karnataka, Lakshadweep, Madhya Pradesh, Maharashtra, Odisha, Punjab, Rajasthan, Tamil Nadu, Uttar Pradesh and West Bengal.

The strata so formed were further sub-stratified. So far as rural areas are concerned, States were categorized into 'hilly' States and other States with 10 States/UTs comprising the former category.[19] If 'r' was the sample size allocated for a rural stratum (hilly States), 'r/2' sub-strata of roughly equal population were formed after arranging the villages within a district as per frame (Census 2001) in ascending order of population. In the other i.e., non-hilly States (excluding Kerala), the villages within a district were arranged in ascending order of proportion of irrigated area in the cultivated area of the village. Then 'r/2' sub-strata were formed comprising a group of villages of the arranged frame such that each sub-stratum had more or less equal cultivated area. In respect of the State of Kerala, the procedure adopted for the hilly states was followed for sub-stratification of FSUs (which were panchayat wards) because of non-availability of information on irrigation at the FSU level. With regard to urban areas, no sub-stratification was attempted for million plus cities. However, for other strata, each district was divided into 2 sub-strata, namely, sub-stratum 1 comprising all towns of the district with population less than 50,000 as per census and sub-stratum 2 consisting of the remaining towns of the district (except million plus cities).

In some of the surveys, each urban stratum has been further stratified to ensure adequate number of ultimate stage units of interest in the sample. For example, in the NSS 69th round (July–December 2012) where urban slum was also a subject of enquiry, each urban stratum (i.e., each city with population of 1 million or more as per Population Census 2011/other cities and taken together within a district) was divided into 2 sub-strata namely all UFS blocks having area type 'slum area' (sub-stratum 1) and the remaining UFS blocks of the stratum (sub-stratum 2). Similarly, in the Household Consumption Expenditure Survey (HCES: 2022–23), each urban basic stratum (i.e., State X Size class of towns: less than 50,000; 50,000 to less than 300,000; 300,000 to less than 1 million; and each city with a population of 1 million or more) has been further divided into two strata, namely, the UFS blocks identified in the sampling frame as belonging to 'affluent area' forming one stratum and the remaining UFS blocks of the basic stratum forming another stratum. This is to ensure adequate representation of affluent households in the sample.

2.12 Sample Size (FSUs)

The decision on sample size in any survey primarily depends on three factors namely (i) the key indicator(s) of interest, (ii) acceptable margin of error of the estimated value of the indicator(s) and (iii) the domain for which such estimates are required. However, quite often the overall cost consideration in terms of resources available and also time frame for completion of the survey has an influence on the plausible overall sample size that can be afforded. The sample size has generally a negative

[19] Hilly States included Arunachal Pradesh, Himachal Pradesh, Jammu & Kashmir, Manipur, Meghalaya, Mizoram, Nagaland, Sikkim, Tripura and Uttarakhand.

relationship with the magnitude of sampling error and positive relationship with the magnitude of non-sampling error in the sense that with the increase in sample size, the sampling error is likely to decline while the non-sampling error is likely to increase. In the NSS, efforts have been to adopt a relatively larger sample size of FSUs with a smaller number of USUs per FSU in view of the fact that between-FSU variability in relation to the study variable is likely to be higher than within-FSU variability. Such a procedure is likely to increase the overall sampling efficiency.

In its efforts to generate improved State-level estimates, the NSS has gradually increased the sample size of FSUs over time by creating additional posts of the field investigating team so as to handle the increased workload. To give some idea, NSS 3rd round, which also covered urban areas for the first time, had a sample size of 920 villages and 490 urban blocks. However, it was a survey of 4 months' duration only (August–November 1951). The earliest round spread over a duration of 12 months was NSS 14th round (July 1958–June 1959), which adopted a sample size of 2,616 villages and 2,216 urban blocks. A significant increase in the sample size took place during NSS 18th round and from this round an overall sample size of about 8,000–9,000 villages and nearly 4,000–5,000 urban blocks have been covered in many NSS rounds. Statement 2.3 gives a rough idea about the trend in sample size over different rounds including the following rounds/surveys: NSS 59th round (being the last round of twelve months' duration after which there has been deeper stratification in urban areas with district as the broad stratum introduced in NSS 61st round), NSS 68th round (i.e., the last quinquennial survey on employment-unemployment and household consumer expenditure), NSS 73rd round on unincorporated non-agricultural enterprises and PLFS, 2022–23. It may be seen that in the recent period NSS 73rd round had the maximum sample allocation of 8,488 villages and 7,860 urban blocks.

2.13 Allocation of Sample Size (FSUs)

Generally speaking, in most of the NSS rounds, overall country level sample size has been allocated over States/UTs, area (rural/urban) and strata following the principle of proportional allocation by considering an appropriate variable associated with the subject of interest. While doing so, some minimum sample allocation to a State/UT x rural/urban x stratum is generally ensured.

Murthy and Roy (1975) give a brief description of the method of overall (i.e., all-India) allocation of sample size over States/strata for the initial NSS rounds.[20] As regards allocation of sample villages, as described by them, 'all-India sample of villages was allocated to the different states on the basis of their rural population (in rounds 1 to 3), total consumer expenditure as estimated from 1st round (in rounds 4 to 7), agricultural population (in round 8), non-agricultural population (in rounds 9 and 10), and again rural population in rounds 11 to 13. Since the fourteenth round,

[20] See Murthy and Roy (1975).

Statement 2.3 Allotted sample size of FSUs in some NSS rounds

NSS round number	Survey period	Subjects of enquiry	Sample size	
			No. of villages	No. of urban blocks
3	August–November 1951	Agriculture and animal husbandry, small-scale manufacture and handicrafts, trade, profession, services and financial operations, consumer expenditure, prices	920	490
14	July 1958–June 1959	Population, births and deaths, employment, land utilization and yield surveys, small scale manufacture and handicrafts, income and expenditure, prices	2,616	2,216
18	February 1963–January 1964	Village statistics, population, births and deaths, migration, housing condition, urban labour force, land utilization, yield survey, earnings from professions and liberal arts, construction, indebtedness of scheduled tribe households in Manipur and Tripura, income of rural labour households, consumer expenditure, prices	8,472	4,572
59	January–December 2003	Situation assessment survey of farmers, land and livestock holdings, debt and Investment, consumer expenditure	6,784	3,824
61	July 2004–June 2005	Household consumer expenditure, employment-unemployment	8,128	4,660
67	July 2010–June 2011	Unincorporated non-agricultural enterprises belonging to manufacturing, trade and other services (excluding construction)	8,380	7,620
68	July 2011–June 2012	Household consumer expenditure, employment-unemployment	7,508	5,276
70	January–December 2013	Situation assessment survey of farmers, land and livestock holdings, debt and Investment	4,529^	3,507^

(continued)

Statement 2.3 (continued)

NSS round number	Survey period	Subjects of enquiry	Sample size	
			No. of villages	No. of urban blocks
73	July 2015–June 2016	Unincorporated non-agricultural enterprises belonging to manufacturing, trade and other services (excluding construction)	8,488	7,860
PLFS	July 2022–June 2023	Employment-unemployment	7,024	5,776

^ Indicates the number of FSUs surveyed. Although NSS 70th round and also NSS 59th round was a survey of one-year duration, the sample size was relatively smaller as compared with the other annual surveys in view of the fact that the survey in these two rounds involved two visits to each sample FSU/household

Sources MoSPI (2001) for NSS round numbers up to 18 and respective NSS reports for the other rounds and PLFS

the allocation to the States is based on a joint consideration of their rural population, geographical area (G), area under cereal crops (GP), and number of investigators'. In respect of allocation of sample blocks, they describe that 'the allocations to the ultimate strata are in proportion to their urban population or non-agricultural population. Since the sixteenth round, double the weight has been given to sector 1 (towns with population 50,000 or more) than to sector 2 (remaining towns), and the all-India sector-wise sample size has been allocated to the States in proportion to respective urban population. However, the double weight to sector 1 towns was discontinued since the twenty-second round….'.

Coming to the recent period, the text that follows describes the method of sample allocation for four surveys covering different subjects, namely, (i) employment-unemployment (as per both NSS 61st round, 2004–2005 and Periodic Labour Force Survey to highlight changes in the method of sample allocation between the two surveys covering the same subject); (ii) unincorporated non-agricultural enterprises (NSS 67th round, 2010–11); and (iii) situation assessment of agricultural households, land and livestock holdings as well as debt and investment (NSS 70th round, 2013). The method of stratification adopted in each of these surveys has already been discussed.

NSS 61st round adopted a total sample size of 12,788 FSUs at the all-India level. This was allocated to different States and UTs in proportion to population as per census 2001. The State/UT level total sample size was further allocated between rural and urban areas in proportion to population as per population census 2001 with 1.5 weightage to urban sector. This procedure resulted in an overall sample size of 8,128 villages and 4,660 urban blocks in the country. For each of the rural and urban areas of a State/UT, the respective sample size was allocated to different strata in proportion to the stratum population as per census 2001. If 'n' was the number of FSUs allocated to a rural/urban stratum, then 'n/2' sub-strata were formed by equalizing population (in case of rural) or number of households (applicable for urban) after arranging the villages/towns of the stratum in ascending order of population as per Census 2001. Finally, a sample of 2 FSUs was allocated to each sub-stratum.

PLFS generally adopts a total sample size of about 12,800 FSUs (about 7,000 villages and nearly 5,800 UFS blocks) annually for the country as a whole. In the PLFS, State/UT level sample size is allocated between rural and urban areas in proportion to population as per Census 2011 with double weightage to urban. Thereafter, State/UT x rural/urban sample size is allocated to different strata and also sub-strata (applicable for rural areas) in proportion to respective population as per Census 2011. Both rural and urban allocations at the stratum level are adjusted to multiples of 8.

NSS 67th round (2010–11) had a sample size of 16,000 FSUs at the all-India level and this was allocated to States/UTs in proportion to non-agricultural workers as per Economic Census (EC)-2005 taking into account the minimum allocations required for a State/UT. State/UT sample size was further allocated to rural and urban sectors in proportion to number of non-agricultural workers as per EC-2005 with a condition that urban allocation should not be too high compared to rural allocation and both rural and urban allocations of the State/UT were adjusted to multiples of 8. Thereafter, State/UT × rural/urban sample size was allocated over strata in proportion to number of non-agricultural workers as per EC-2005. For the States/UTs where census 2001 frame was used in the rural sector, allocations to strata were made in proportion to population as per census. Finally, stratum level sample size was allocated over sub-strata in proportion to number of non-agricultural workers as per EC-2005 in rural sector as well as in million plus cities (after assuming the number as 1 for those villages/blocks where there was no non-agricultural worker), and in proportion to number of blocks with a double weight to sub-stratum 1 for other than million plus cities. The minimum allocation for a sub-stratum was 4.

So far as NSS 70th round (2013) is concerned, the total sample size of 8,042 FSUs was allocated to States/UTs in proportion to population as per census 2011 with a minimum allocation to each State/UT. State/ UT level sample size was allocated between rural and urban areas in proportion to population as per census 2011 with double weightage to urban subject to the restriction that urban sample size for bigger states like Maharashtra, Tamil Nadu etc. did not exceed the rural sample size. Thereafter, State/ UT x rural/urban sample size was allocated to different strata in proportion to population as per census 2011. Allocations at stratum level were adjusted to multiples of 2. For special stratum formed in the rural areas of 20 States/ UTs, a sample of 2 FSUs was allocated to each. For special stratum 1 in the rural areas of Nagaland and Andaman & Nicobar Islands, 4 and 2 FSUs were allocated respectively. Allocation was 2 for each sub-stratum in the rural areas. In urban, stratum level sample size was allocated between the two sub-strata in proportion to number of FSUs with a minimum allocation of 2 FSUs to each sub-stratum.

2.14 Method of Sampling of FSUs

The choice of the exact method of sampling in any sample survey primarily depends upon the type of auxiliary information available in respect of each unit in the sampling frame and the subject(s) of enquiry. Since its introduction in 1950, for the sampling of FSUs (which are mostly villages in rural areas and urban blocks in urban areas), the NSS has adopted mainly four different sampling procedures, namely, probability proportional to size (PPS) with replacement, circular systematic sampling with equal probability, circular systematic sampling with probability proportional to size, and simple random sampling without replacement.

Murthy and Roy (1975), in their article on 'Development of the Sample Design of the Indian National Sample Survey during its First Twenty-Five Rounds', clearly describe different sampling procedures adopted in these early rounds of the NSS for sampling of villages and blocks. The most relevant portions from their article are reproduced.[21]

(a) Sampling of villages: 'In the first three rounds, when the 1941 census list of villages was used as the sampling frame, villages, the first stage units, were selected with PPS with replacement, size being village population (whenever available) or village area (if population figures were not available). Simple random sampling was used for areas for which neither population nor area figures were available……In rounds 4 to 7, tehsils were selected as first-stage units and villages as second-stage. The reason for this departure was to make use of the 1951 census frame. Collection of the whole frame of nearly 600,000 villages would have meant a delay of 2 or 3 years more. Hence, it was decided to select a sample of tehsils in the first instance, to collect the village frame (by copying) only for the sampled tehsils and select villages from these frames. Tehsils were sampled with PPS with replacement (size being population in most cases and area in others) and the same scheme was adopted for villages also. With the arrival of the complete 1951 census frame in 1954, switch-over was made to the direct sampling of villages in the eighth round. This has continued since then. During eighth to thirteenth rounds, villages meant for both household enquiries and crop surveys were selected with probability to proportional to population (ppp) and with replacement whereas in the tenth to thirteenth rounds additional sample villages, intended for crop survey only, were selected with probability proportional to village area (ppa) and with replacement….A major change in the fourteenth round was the re-introduction [introduction] of equal probability sampling. This was done in order to effect full and economic integration of socio-economic enquiries and crop surveys by having just a common set of villages for them without any additional sample for the latter…..After having opted for equal probability sampling, it was decided to have [circular] systematic sampling of villages in preference to simple random sampling since the former – with existing arrangement in the frame or with rearrangement – provides a more

[21] This section contains excerpts from Murthy and Roy (1975).

representative sample....This procedure continued up to the seventeenth round. It was in the eighteenth round that PPS sampling of villages was again used in the NSS. The size measure was the village population rounded off to the next higher multiple of 1000.....Village sizes were revised in the nineteenth round when the size of a village was obtained as the ratio of the village population to the average population of the villages in the population class 0-499 in that state, rounded off to a suitable integer. However, sampling was done – not with replacement – but in a circular systematic manner with PPS'.[22]

(b) Sampling of urban blocks: 'Since a complete frame of blocks was not available during the third to eighth rounds, at first towns were sampled and then blocks were selected from them, after collecting the lists of blocks for the sample towns. Selections at both the stages were with probability proportional to population and with replacement. Since the ninth round town selection was given up as the complete list of blocks became available by this time. Then blocks were selected circular systematically with equal probability up to the seventeenth round, and with probability proportional to size (related to block population) thereafter, the size of a block being its population in thousand rounded off to the next integer'.[23]

The MoSPI's Golden Jubilee publication (2001) titled 'Concepts and Definitions Used in NSS' summarizes the sampling procedures of FSUs adopted till NSS 55th round (1999–2000). A cursory look at the same reveals that among the four sampling procedures followed in the NSS (as described at the first paragraph of this section), PPS with replacement sampling was the most common method used during this period for sampling of villages. It was used in 30 out of 55 NSS rounds. This method was followed by circular systematic sampling with PPS, which was used during 17 NSS rounds. On the other hand, for sampling of blocks, the most commonly used method was circular systematic sampling with equal probability, which was used in 22 NSS rounds. The pattern also remained similar during the initial period covering first twenty-five rounds and the subsequent period till the 55th round.

Turning back to the NSS rounds/surveys till today subsequent to NSS 55th round, again for the rural areas, the most commonly used method has been the PPS with replacement for sampling of villages. However, for the urban areas, simple random sampling without replacement has been found to be the most frequent one for sampling of urban blocks. It is of interest to note that the method of circular systematic sampling (with PPS in case of selection of villages and equal probability in case of selection of urban blocks) which was used for a considerable period till NSS 57th round (2001–2002) has been given up since NSS 58th round (July–December 2002). Instead, PPS with replacement procedure for sampling of villages and simple random sampling without replacement for sampling of urban blocks have been used with the exception being NSS 65th round (2008–2009) which adopted circular systematic sampling as was the practice prior to NSS 58th round.

[22] Murthy and Roy (1975).

[23] Ibid.

For each of the sampling procedures, Box 2.2 mentions the corresponding NSS round numbers/other recently conducted surveys where the particular method of sampling was used. It may be seen from Box 2.2 that five different sampling schemes have been used in different NSS rounds and the other surveys. These include probability proportional to size with replacement (PPSWR), Circular systematic sampling with equal probability, Circular systematic sampling with probability proportional to size (PPS), Simple random sampling without replacement (SRSWOR) and Simple random sampling with replacement (SRSWR). In the PPSWR method of sampling, the FSUs are selected with a 'size measure' attached to each FSU which is either population or number of households or number of enterprises/workers (applicable to enterprise surveys) in the FSU as per the sampling frame in most of the surveys. Such a method leads to the selection of larger FSUs (in terms of the FSU size) in a much higher proportion in the overall sample. This method is quite often suitable in cases where the FSUs vary significantly in terms of their sizes. The method of circular systematic sampling (with equal probability or with PPS) used in the NSS involved the arrangement of the FSUs in geographical order as per the sampling frame within each first-stage stratum. This method ensures a wider geographical spread of the sample FSUs within the first-stage stratum. The method of SRSWOR/SRSWR gives equal chance of selection to each FSU in the sampling frame irrespective of their sizes i.e., whether they are large or small.

2.15 Use of Independent Sub-Samples

A salient feature of the NSS has been to draw the required number of sample FSUs from any stratum in the form of independent sub-samples, generally two sub-samples, with sample FSUs for each sub-sample drawn as per the same method of sampling. With this procedure, it is possible to generate independent estimates of aggregates and ratios of interest based on the sample FSUs and USUs covered under each sub-sample. One major advantage of this procedure is its simplicity to estimate the sampling variance and relative standard error (RSE) of the estimates of aggregates and ratios even when the sample design including the method of sampling is a complex one.

To illustrate the above, if sample FSUs from any stratum 's' are drawn in the form of 'm' independent interpenetrating sub-samples of same size according to the same sampling procedure and if $\widehat{Y}_{si}$ is the unbiased estimator of the total of the study variable for s-th stratum based on i-th sub-sample, then an unbiased estimator of the sampling variance of total of s-th stratum $\left(\hat{Y}_s\right)$ based on all the m sub-samples is given by[24]

[24] See Murthy (1967).

Box 2.2 Method of sampling of FSUs used in different NSS rounds

Sl. No.	Method of sampling	NSS rounds	
		Rural	Urban
1	Probability proportional to size (PPS) with replacement	1–13, 28–40, 42, 48, 49, 51, 58–64, 66–69, 71–73, 75, PLFS Series till December 2024, 76	3–8, 28–34, 40, 51, 71–73£, 75, PLFS Series till December 2024
2	Circular systematic sampling with equal probability	14–17, 25–27, 40, 45, 46, 52, 54, 74#,	9–17, 35, 40, 41, 43–47, 50, 52–56, 65, 74#,
3	Circular systematic sampling with probability proportional to size (PPS)	18–24, 41, 43–47, 50, 53, 55, 56, 57*, 65	18–27, 41, 45, 46, 56, 57*, 62^, 63^, 67**,
4	Simple random sampling without replacement	56, 70, 77, TUS: 2019, 78, ASUSE: 2021–22, 79, HCES: 2022–23, ASUSE: 2022–23, HCES: 2023–24, ASUSE: 2023–24, TUS: 2024, ASUSE: 2025, 80, PLFS: 2025	36–39, 42, 48, 49, 56, 58–61, 62!, 63!, 64, 66, 67@, 68–70, 73£, 77, TUS: 2019, 78, ASUSE: 2021–22, 79, HCES: 2022–23, ASUSE: 2022–23, HCES: 2023–24, ASUSE: 2023–24, TUS: 2024, ASUSE: 2025, 80, PLFS: 2025
5	Simple random sampling with replacement	76***	76

[TUS: Time Use Survey; HCES: Household Consumption Expenditure Survey; ASUSE: Annual Survey of Unincorporated Sector Enterprises]

(Different methods of sampling were used in various ultimate strata in some of the NSS rounds devoted to enterprise survey like round number 40, 41, 45, 46, 56, 62, 63, 67 and 73. In NSS 76th round also, for the rural areas, a mix of PPS with replacement sampling and simple random sampling with replacement was used)

*Used for most of the strata

^ Used for 27 cities with population of one million and above

**Used for 26 cities with population of one million and above (which excluded Mumbai)

! Used in respect of cities and towns other than 27 cities with million plus population

@ Used for the city of Mumbai and also in respect of cities and towns other than the remaining 26 cities with million plus population

£ NSS 73rd round, being devoted to the enterprise survey, primarily used the sixth Economic Census information on number of enterprises and workers at the village/block level for stratification and sampling of FSUs. However, for the urban areas of 15 States/UTs (applicable for first six months of the surveys), Economic Census frame could not be used for this purpose and instead UFS frame was used with sampling of blocks by simple random sampling without replacement

NSS 74th round was a list frame-based survey of bigger enterprises with the list/sampling frame developed based on three sources namely Sixth Economic Census, Business Registers prepared by States and the list of companies as per the database of MCA (M/o Corporate Affairs). The survey adopted a uni-stage sampling

*** In the NSS 76th round, PPS with replacement and simple random sampling with replacement procedures were followed for sampling of FSUs in different rural strata

$$\hat{V}(\hat{Y}_s) = \frac{1}{m(m-1)} \sum_{i=1}^{m} (\hat{Y}_{si} - \hat{Y}_s)^2$$

where $\hat{Y}_s = \frac{1}{m} \sum_{i=1}^{m} \hat{Y}_{si}$

In case the sample FSUs from each stratum are drawn in the form of 2 independent sub-samples i.e., m = 2 (which is the case with the NSS rounds), then the above expression of the estimator of the sampling variance of $\hat{Y}_s$ gets converted to:

$$\hat{V}(\hat{Y}_s) = \frac{1}{4} (\hat{Y}_{s1} - \hat{Y}_{s2})^2$$

where suffix 1 stands for sub-sample 1 and suffix 2 for sub-sample 2.

Accordingly, the sampling variance of the estimate of aggregate Y at the State/ UT/ all-India level (for rural or urban as the case might be) can be estimated by adding the stratum level variances as follows:

$$\hat{V}(\hat{Y}) = \frac{1}{4} \sum_{s=1}^{K} (\hat{Y}_{s1} - \hat{Y}_{s2})^2$$

where K denotes the total number of ultimate strata formed within the State/UT (rural or urban).

Finally, the RSE of $\hat{Y}$ may be estimated as

$$\widehat{RSE}(\hat{Y}) = \frac{\sqrt{\hat{V}(\hat{Y})}}{\hat{Y}} X\, 100$$

Similarly, if R is the State/UT/All-India level ratio of two aggregates Y and X i.e., R = Y / X, which can be estimated as $\hat{R} = \hat{Y} / \hat{X}$, the RSE of $\hat{R}$ can be estimated as.

$\widehat{RSE}(\hat{R}) = \frac{\sqrt{\widehat{MSE}(\hat{R})}}{\hat{R}} X\, 100$ where

$$\widehat{MSE}(\hat{R}) = \frac{1}{4\hat{X}^2} \sum_{s} \left[(\hat{Y}_{s1} - \hat{Y}_{s2})^2 - 2\hat{R}(\hat{Y}_{s1} - \hat{Y}_{s2})(\hat{X}_{s1} - \hat{X}_{s2}) + \hat{R}^2 (\hat{X}_{s1} - \hat{X}_{s2})^2 \right]$$

Apart from its simplicity to measure RSE of estimates, this method of use of independent and parallel samples, termed as interpenetrating sub-samples developed by Professor Mahalanobis, has other advantages as well. For example, the simple divergence between the two (or more) sub-sample estimates is likely to give a rough idea about the level of precision of the estimate based on the combined sub-sample. Similarly, if primary data based on the two sub-samples is collected by two different parties for a reasonable period of time and if there is a pattern in the observed divergence between the alternative estimates based on the two sub-samples, it can reflect the relative data quality aspects of the two parties. In the earlier NSS rounds, as the Concepts and Definitions Used in NSS, MoSPI (2001) mention, 'different parallel samples were surveyed by two parties of investigators and the data were processed in different processing centres to help in providing a broad check on the survey results and analysing the total error into its different components'. This technique has been given up since 1972 'to have economy in the collection and processing costs' (MoSPI, 2001).

2.16 Segmentation of Large FSUs and Selection of Segments

The preparation of the sampling frame of households or enterprises to draw a representative sample for detailed enquiry involves a listing exercise in the selected FSUs. In case the sample FSU is large in size in terms of population (and/or number of enterprises in case of enterprise survey), the said listing exercise involves a huge amount of time. To overcome this problem, the general practice in the NSS from its 14th round (July 1958–June, 1959) has been to divide the large villages/urban blocks having the approximate present population of 1,200 or more[25] into a certain number of segments called 'hamlet-groups' (HGs) in rural areas and 'sub-blocks' (SBs) in case of urban by roughly equalizing population, and finally selecting a random sample of such segments (one or two in number) for the listing purposes.[26]

The larger is the village/urban block in terms of approximate present population (P), the higher is the total number of segments (D) formed. In the enterprise surveys where all enterprises within the FSU are also listed, apart from the present population, approximate total number of enterprises (E) in the FSU is also considered to decide the total number of segments to be formed.

The guidelines given for deciding the value of D in the PLFS: 2022–23 (household survey)—with this method being prevalent for a long time—and in the NSS 73rd round: 2015–16 (enterprise survey) are summarized in Box 2.3.

The criteria adopted for HG/SB formation are equality of population content[27] and geographical contiguity with a priority given to geographical contiguity in case there

[25] 600 or more in respect of rural areas of Himachal Pradesh, Sikkim, Nagaland and Poonch, Rajouri, Udhampur, and Doda districts of Jammu and Kashmir. Over time the coverage of these areas where this lower threshold of 600 or more population is used to form HGs has been extended. In NSS 73rd round (2015–16), such areas included rural areas of Himachal Pradesh; Sikkim; Andaman & Nicobar Islands; Uttarakhand (except four districts Dehradun, Nainital, Hardwar and Udham Singh Nagar); and the districts of Punch, Rajouri, Udhampur, Reasi, Doda, Kishtwar, Ramban, Leh (Ladakh), Kargil and Idukki district of Kerala. This procedure also continued in the PLFS: 2023–24.

[26] Such a procedure of segment selection from large villages for the purpose of listing of households was introduced for the first time in NSS 9th round (May–November 1955). However, during 9th round to 13th round, the general practice was to select one hamlet of the large village with probability proportional to population.

[27] The criterion of HG/SB formation is based on the equality of number of enterprises in case of many enterprise surveys including the NSS 73rd round survey if the decision on the exact number of

Box 2.3 Number of HGs/SBs ('D') formed in an FSU

PLFS: 2022–23		NSS 73rd round: 2015–16*			
Value of P	D	Value of P	D	Value of E	D
<1,200	1	<1,200	1	<120	1
1,200–1,799	3	1,200–1,599	4	120–159	4
1,800–2,399	4	1,600–1,999	5	160–199	5
2,400–2,999	5	2,000–2,399	6	200–239	6
(and so on)	..	(and so on)	..	(and so on)	..

*Final value of D in NSS 73rd round was the higher of the two values based on P and E

is a conflict between the two. For rural areas/villages, HG formation involves the listing of natural hamlets of the village from the north-west corner of the village and proceeding southwards in a serpentine manner and numbering them in that order. If there is no natural hamlet, census sub-divisions of the village like enumeration blocks or groups of census house numbers or geographically distinct blocks of houses are treated as 'hamlets'. HGs are thereafter formed by including contiguous hamlets under a HG so that they have roughly equal population. Large hamlets are divided artificially to achieve equality of population content of each HG. Care is taken to see that there is not substantial difference between the population (or number of enterprises as the case might be) of the smallest HG and the largest HG formed. For the numbering of HGs, the HG containing hamlet number 1 is numbered as 1, the HG with the next higher hamlet number not included in HG 1 is numbered as 2 and so on. The procedure for formation of SBs is also similar.

In the PLFS: 2022–23, out of all the HGs/SBs formed in the FSU, the one having maximum share in the total population of the FSU was selected with certainty and from among the remaining one more was selected by simple random sampling for listing purposes. The listing and sampling of households in the two selected HGs/SBs were done independently. In NSS 73rd round, the HG/SB having the maximum number of enterprises/establishments under the coverage was selected with certainty, termed as 'segment 1'. Two more HGs/SBs were selected from the remaining ones by simple random sampling without replacement and they were combined to form 'segment 2'. Listing and selection of the enterprises were done independently in segments 1 and 2.

As regards the history of segmentation of FSUs in the NSS, the following excerpts from Murthy and Roy (1975) may be useful for the readers: 'In order to reduce the investigator's work load in a large sample village the technique of hamlet selection was introduced in the ninth round (1955). A large village usually contains some recognized sub-divisions (called hamlets) having names of their own. One of the hamlets in a village was selected in the following manner: (i) The hamlets were listed in the alphabetical order of their names; (ii) For each hamlet, the percentage of

HGs/SBs formed the FSU is guided by the number of enterprises criterion rather than the population criterion.

village population contained in it was ascertained by local enquiry or eye-estimate and recorded; (iii) The percentage figures were cumulated and one hamlet was selected with probability proportional to percentage value; and (iv) The survey was confined to the selected hamlet only. But this procedure was not found to be desirable, since sometimes hamlet selection was resorted to just because the sample village had some hamlets. Secondly, there was possibly a tendency for the field staff to select smaller hamlets more often and lastly the estimated percentage figures were too approximate to adopt PPS selection. So in the fourteenth round (1958–59) the procedure of hamlet selection was modified as follows: (i) The number of sub-divisions to be formed in a village was pre-specified on the basis of the last census village population figures and the expected growth rates; (ii) The specified number of sub-divisions were formed by grouping contiguous hamlets so that the sub-divisions (called hamlet-groups) were compact areas having more or less equal population content; (iii) An objective rule was specified to refer to random number tables for selecting one of these groups at random with equal probability; and (iv) The survey was confined to the selected group only. A similar procedure is followed in the urban areas also'.

2.17 Sampling Frame of Ultimate Stage Units

The households or enterprises, as the case might be, constitute the ultimate-stage units (USUs) in the NSS. The sampling frame of USUs i.e., the list of households/ enterprises as on the date of survey is prepared by the field investigators in the selected villages/urban blocks by carrying out a listing exercise which involves visit to every structure/house/household (or enterprise) in the sample FSU (village/urban block) from the north-west corner. For listing purposes, the large FSUs with approximate present population exceeding or equal to 1,200 (the threshold being 600 or more in case of certain hilly areas as already mentioned) are divided into a few segments (HGs/SBs) and the listing exercise for preparing the frame of USUs is confined to selected segment(s) only.

2.18 Stratification of USUs

At the time of listing some auxiliary information related to the subject of enquiry is collected in the NSS in respect of each USU for stratifying the USUs before sampling. The aim of the stratification is to ensure adequate representation of different types of households/enterprises in the sample for improving the efficiency of the estimates. While doing so efforts are also made to identify households/enterprises which may be relatively rare in number but their overall contribution with regard to a parameter of interest may be significant. For example, it may be useful to form strata comprising relatively rich households as well as poorer households in the surveys on household consumer expenditure; strata of households having significant variations in terms of land possessed in the surveys on land holding/agricultural households; strata of households having members with different type of disabilities in the surveys on disabled persons; and similarly, forming strata of enterprises belonging to different types of industries in joint consideration with employment size in the case of enterprise surveys.

To give some specific examples, in the last 'quinquennial survey' i.e., NSS 68th round (2011–12) survey on employment-unemployment and household consumer expenditure, listed households were stratified into 3 strata. For rural areas/villages, the three strata were: relatively affluent households (stratum 1); of the remaining, households having principal earning from non- agricultural activity (stratum 2); and all remaining households (stratum 3). For the urban areas/sample UFS blocks, the auxiliary information i.e., the stratification variable was the household monthly per capita consumer expenditure (MPCE) and the listed households were grouped into 3 strata: households having MPCE equivalent to top 10% of urban population (stratum 1); households having MPCE corresponding to the middle 60% of urban population (stratum 2); and households having MPCE of bottom 30% of urban population (stratum 3).

As against the above, PLFS considers the number of members (say, M) in the household with secondary or above level of general education as the stratification variable for stratifying the households in each FSU. While 3 strata are formed in each rural FSU (village), the number of strata is 4 in case of each urban sample FSU (i.e., UFS block) as follows: (a) Rural: households with $M = 2$ or more (stratum 1), households with $M = 1$ (stratum 2), households with $M = 0$ (stratum 3); and (b) Urban: households with $M = 3$ or more (stratum 1), households with $M = 2$ (stratum 2), households with $M = 1$ (stratum 3), and households with $M = 0$ (stratum 4).

In the situation assessment survey of agricultural households in NSS 77th round (2019), the listed households were stratified into 5 strata, namely, the non-agricultural households forming one stratum while the agricultural households were stratified into 4 strata based on the area of land possessed by the household (say, A in hectare) as the auxiliary information: households with A less than 0.25 ha; households with A = 0.25 ha–1 ha; households with A = 1 ha–2 ha; and households with A being 2 ha or more.

Similarly, in NSS 73rd round enterprise survey (2015–16), listed enterprises in each FSU/selected segment were stratified into 16 strata by jointly considering (a) broad activity of the enterprise (i.e., manufacturing, trade and other services), (b) whether the enterprise normally hired a worker or not, and (c) industry (National Industrial Classification) code of the enterprise.

2.19 Number of Sample USUs Per FSU

As regards the allotted sample size i.e., number of USUs (households/enterprises) per FSU (village/urban block), the practice has been to keep it reasonably smaller but not too small since it may not be cost-effective in view of (a) considerable travel

Box 2.4 Allotted number of households/enterprises per FSU

Subject	NSS round (survey period)	Sample size per FSU
Household Consumer Expenditure	68 (2011–12)	8
Household Consumer Expenditure	HCES (2022–23)	18
Employment-unemployment	68 (2011–12)	8
Employment-unemployment	PLFS (2017–18 & onward)	8 (12 in PLFS: 2025)
Non-agricultural Enterprises	73 (2015–16)	32*
Land & Livestock Holdings	77 (2019)	10
Agricultural Households	77 (2019)	10
Debt & Investment	77 (2019)	12
Time Use	TUS (2024)	14

*In addition, all 'big enterprises' in the sample FSU with 20 or more workers were surveyed. Allotted sample size per FSU was somewhat larger since there was a deeper stratification to ensure representation of enterprises belonging to different industrial categories

time (particularly in the rural areas) to reach the FSU and consequently the cost involved in it and (b) the time required to carry out the listing exercise for preparing the frame of USUs. Box 2.4 indicates the allotted number of USUs per FSU in the latest surveys on a few important subjects of enquiry.

2.20 Sampling Procedure of USUs

Generally, the practice has been to select the required number of USUs from any stratum by simple random sampling without replacement. However, prior to NSS 57th round (2001–2002), in many rounds the USUs were selected by circular systematic sampling (with equal probability) after arranging the USUs as per a variable likely to be correlated with the main subject of enquiry. For example, in the NSS 32nd round (1977–78) survey i.e., the second quinquennial survey on employment-unemployment and household consumer expenditure, all the listed households of a sample village were arranged according to their means of livelihood namely self-employed in non-agriculture, rural labour and 'other households' as per the main source of income of the households during the last 365 days. In the sample urban blocks, the households were arranged into four classes based on their nature of employment (self-employed, not self-employed) and their per capita expenditure level (less or greater than the pre-specified per capita expenditure for the State). Similarly, in the NSS 56th round (2000–2001) survey on unorganized manufacturing enterprises, after 3 strata were formed respectively comprising directory manufacturing establishments (DMEs i.e., units with 6 more workers including at least one hired worker), non-directory manufacturing establishments (NDMEs i.e., units less than 6 workers including hired workers) and own account manufacturing enterprises (OAMEs i.e., the manufacturing units generally working without any hired workers), the required number of units from each stratum was selected by circular systematic sampling after arranging them according to 'Broad Manufacturing Group' codes. In fact, this was the last NSS round after which the method of circular systematic sampling of USUs has been given up and instead simple random sampling without replacement method is followed.

2.21 Self-Weighting Design

For quite a few NSS rounds since NSS 4th round, NSS used the 'self-weighting' design which simplified tabulation exercise. However, this technique has been given up after experimentation in quite a few NSS rounds. One disadvantage of this design was that it resulted in unequal number of USUs per FSU causing practical difficulties.

In a self-weighting design, the survey weights i.e., 'multipliers' for all sample households of a given domain of interest (like State) become constant (say, K) and accordingly the population total can be easily estimated by simply summing up the sample values of all the households and multiplying that sum total by K.

To illustrate, if sample FSUs are drawn from a stratum by probability proportional to population (as per previous census with its list of FSUs used as the sampling frame of FSUs) with replacement and the required number of sample households in the FSU are selected with equal probability from among the listed households, then State total (for rural or urban as the case might be) for any characteristic of interest 'y' is estimated as

$$\hat{Y} = \sum_{s} \frac{P_s}{n_s} \sum_{i=1}^{n_s} \frac{1}{p_{si}} \frac{H_{si}}{h_{si}} \sum_{k=1}^{H_{si}} y_{sik}$$

where s = stratum, P_s = total census population of s-th stratum, n_s = number of surveyed FSUs of s-th stratum, 'i' stands for sample FSU, p_{si} = census population of i-th village of s-th stratum, 'H' and 'h' respectively denote number of households listed and number surveyed, and suffix 'k' stands for sample household.

In a self-weighting design, the multiplier attached to the sample household is constant 'K' i.e.

$$\frac{P_s}{n_s} \cdot \frac{1}{p_{si}} \cdot \frac{H_{si}}{h_{si}} = K \quad (\text{with} \frac{H_{si}}{h_{si}} = \frac{K.n_s.p_{si}}{P_s})$$

where K can be approximated as the ratio of projected number of households and targeted number of sample households.

Accordingly, State total can be estimated as $\hat{Y} = K \sum_{s} \sum_{i=1}^{n_s} \sum_{k=1}^{H_{si}} y_{sik}$.

In practice, the self-weighting design is achieved by specifying the values of sampling fraction of households (H_{si}/h_{si}) for each FSU to the field investigators.

2.22 Method of Data Collection

The field investigators deputed for fieldwork visit the selected USUs i.e., households or enterprises depending upon the subject of enquiry and collect the required information as per the detailed schedule of enquiry by interviewing the most appropriate respondent of the household/enterprise who is in a position to furnish the bulk of the information. Thus, the method of data collection is based on the oral enquiry. However, in case of enterprise surveys, if the selected enterprise maintains books of accounts, efforts are made to collect the requisite information as per the records/ books of accounts.

2.23 Substitution of Samples

Whenever it is difficult to survey any selected USU (or the sample FSU), it is substituted by another sample similar to the originally selected one from the same stratum. In case of USU, there is a provision to go for substitution twice after which the sample is treated as a casualty in case the substituted USU does not provide the information or if it cannot be contacted.

2.24 Uniform Spread of Fieldwork and the Concept of Sub-Round

Most of the NSS rounds and other surveys conducted by the NSSO are of one-year duration. There are some rounds with a survey period of six months as well. The surveys of one-year duration are divided into four 'sub-rounds' with a duration of three months each. Similarly, if the survey period is of six months' duration, it is divided into two sub-rounds, each spanning over a period three months. Equal number sample FSUs is allotted to each sub-round. Any sample FSU allotted to a particular sub-round is surveyed during the sub-round period to which it is allotted. For the samples allotted to a particular sub-round, efforts are also made to survey them in more or less equal number in each of the three months of the sub-round. This procedure ensures uniform spread of fieldwork across different months of the survey period and accordingly helps eliminate the effect of seasonality, if any, in the study variables on the final estimates based on the full sample.

2.25 Moving Reference Period

In view of the uniform spread of fieldwork over the entire survey period, the reference period for a given variable also becomes a moving one. The reference periods generally vary as per the variables of interest. For example, the reference period is the date of survey for variables like land possessed by the household or educational qualification or marital status of a household member or the industry of an enterprise. The reference period is the last week preceding the date of survey (in respect of the 'current weekly status' of a household member in the employment-unemployment surveys/PLFS) or last 30 days (in case of household consumer expenditure on most of the food items) or even last 365 days (for household consumer expenditure on durable goods, for example). The moving reference period helps in reducing the recall error as well as in controlling the seasonal effect, if any.

2.26 State Participation

Another distinguishing feature of the NSS is the participation of most of the State/UT Governments in the survey programme where they cover an independent matching sample of FSUs which are selected following the same sampling design. While the sample FSUs covered by the staff of NSSO or under the direct supervision of NSSO are known as 'central sample', the FSUs covered by the State/UT Governments are described as 'state sample'.

In fact, in NSS 8th round (July 1954–March 1955) when the subject of land holding was taken up as part of the World Agricultural Census, the State Governments were requested to participate in the NSS programme so that it was possible to generate state-level estimates by pooling the two data sets with increased sample size. This was the first NSS round with State participation in the NSS programme and since then it has been a regular feature. As of now almost all the States/UTs barring only a few are participating in the NSS. In NSS 78th round (January–December 2020) all the States and Union Territories except Dadra & Nagar Haveli and Lakshadweep participated with the following matching pattern[28]: Nagaland (Urban)—triple matching; Manipur and Telangana—double matching; Maharashtra (Urban)—one and half times matching; and Remaining States/UTs—equal matching.

With the State Governments' participation in the NSS it is possible to have two independent estimates of any survey characteristic with one based on the central sample and the other as per the state sample. The magnitude of divergence between the estimates based on the central and state samples is likely to give some idea about the level of precision of the estimate assuming that the overall magnitude of the non-sampling error is of the similar order as per the two sources. More importantly, it is possible to pool the two data sets (with consequent increase in the overall sample

[28] MoSPI (2023).

size) for improved estimate at the level of State/UT and also the estimates at the sub-State level i.e., by NSS region or by district.

2.27 Involvement of the Technical Bodies

An important feature of the NSS is that any survey is launched only after the due approval of the survey methodologies by an apex technical body. Earlier, since the formation of the NSSO, this role was vested with the Governing Council of NSSO having total independence in the matter of taking decisions on the subjects of enquiry in a given NSS round and in the method of data collection, data processing and publication of the NSS data. Apart from a non-official Chairman, the Governing Council had five academicians, five data users from Central and State Government departments and senior officers of the Government as its members. The Director General and Chief Executive Officer of NSSO acted as the Member-Secretary of the Governing Council. With the formation of the National Statistical Commission (NSC) in July 2006 as per the recommendation of the Rangarajan Commission, the role used to perform by the Governing Council was subsumed within the functions of the NSC.[29]

The draft survey instruments, particularly the sample design and schedules of enquiry of any NSS round, are first discussed and deliberated at length in the meetings of the Working Group/Standing Committee comprising eminent experts in the related subjects and Government officials. Only after their clearance of the survey methodology and with the concurrence of the same by the National Statistical Commission, the fieldwork for any NSS round is launched.

2.28 Achievements of the NSS

In the initial years after the creation of the NSS, while the work of data collection was handled by the Directorate of NSS in the Government of India, all other technical aspects of the survey work including the preparation of sample design, designing of the questionnaires/schedules of enquiry, the work of data processing and writing

[29] The NSC has a part-time Chairperson and four part-time Members with the Chief Executive Officer, NITI Aayog (initially, Secretary, Planning Commission) as an ex-officio Member and the Chief Statistician of India / Secretary, MoSPI as the Member Secretary of the NSC. To quote from the Annual Report, 2007–2008 of the NSC (see Section 5.3 of the Report), the NSC 'recommended to the Government that since the NSC had come into existence as an independent professional body with requisite autonomy and has the requisite authority as per the Resolution of the Government … and the NSSO has been brought under its umbrella, the Governing Council of NSSO would not be needed and that a Steering Committee reporting to the Commission would be formed to take over the functions performed by the Governing Council of the NSSO. This recommendation of the NSC was accepted by the Central Government, and the Governing Council of NSSO was dissolved' (MoSPI, 2008).

of reports were taken care of by the Indian Statistical Institute under the general direction and guidance of Professor P. C. Mahalanobis. Since its early rounds, the scholars like D. B. Lahiri, J. M. Sengupta and M. N. Murthy were associated with the NSS. Their guidance and unmatched contributions in the initial rounds of the NSS towards development of the sample design shaped the very key features of the survey methodology of the NSS. As narrated by Ghosh, Maiti, Rao and Sinha,[30] according to D. B. Lahiri who is one of the chief architects of the NSS, 'the NSS chose a multi-purpose and multi-subject frame work, and the survey design was gradually so evolved as to permit study of the inter-connections between the various components of the socioeconomic picture of the country and its constituent regions and its states .....'.

One of the greatest achievements of the NSS is the availability of its unit-level (i.e., household and establishment[31] level) data for research and other analysis. This effort started with sharing of data at a very nominal price to begin with. Quite often those were shared free of cost to various institutes for research purposes with properly signed Memorandum of Understanding. Subsequently, the condition of paying even the nominal price was waived. Now anyone can access the freely downloadable data easily from the website of the MoSPI. The access of unit-level data has helped the data users in undertaking various research activities. It has also contributed to a large number of articles of academic interest.

Another accomplishment is the uploading of the softcopy of NSS reports based on various rounds and reports on other socio-economic surveys conducted by the NSSO on the website of the MoSPI. These reports serve as useful reference documents not only in studying the key findings of the surveys but also in planning any survey on a specific subject if one is so interested.

The conduct of surveys at certain intervals on some subjects of enquiry like non-agricultural enterprises, employment-unemployment and household consumer expenditure has helped in analysing the trends in the key indicators based on such surveys. For example, the data based on quinquennial series of the survey on household consumer expenditure (HCE) has helped in analysing how the incidence of poverty has declined over time and what has been the trend in inequality so far as the level of living is concerned in terms of HCE.

The data on certain subjects as per the NSS can also be used to crosscheck the quality of data from other sources. For example, the information on number of establishments as per the Economic Censuses (EC) can be compared with the NSS estimate of number of establishments engaged in manufacturing, trading and other services sector to assess the extent of under-listing in the EC and take corrective measures. Similarly, the number of manufacturing establishments with 10 or more workers as per the NSS enterprise surveys can give a rough idea about the inadequacy of the list

[30] Ghosh et al. (1999).

[31] Although generally described in the NSS as the enterprise surveys, such surveys consider 'establishment' i.e., a unit located in a single physical location as the ultimate sampling unit.

of factories covered under the Factories Act, 1948 which serves as a sampling frame for the Annual Survey of Industries.[32]

On the methodological front, the Pilot Survey on Income, Consumption and Savings (September 1983–December 1984) revealed a number of interesting findings regarding the quality of respective data (see Chapter 5 for details). The findings can be effectively utilized to conduct further surveys on the subjects. Similarly, the Pilot Survey on Suitability of Different Reference Periods for Measuring Household Consumption (January–June 2000) came out with specific suggestions (Chapter 5) regarding the ideal reference period(s) for different items of household consumption and those have been duly taken into consideration in the latest surveys on household consumption expenditure.

NSS has already completed its 74 years of operation and it is the time to celebrate its 75 years of journey. Undoubtedly, the NSS has generated a wealth of information on a wide range of subjects which are used by the planners, policymakers, researchers and other data users. In the words of Deming (1973), 'No country, developed, underdeveloped or over-developed, has such a wealth of information about its people as India has.....' (Rao, 2012).

References

Deming, W. E. (1973). P.C. Mahalanobis—A Profile. *Commemoration Brochure issued on the occusion of the 80th Birth Anniversary of PC. Mahalanobis.* ISI Calcutta: Kolkata.

Department of Statistics. (1985). *Report on the third quinquennial survey on consumer expenditure, 38th round, January–December 1983, NSS report number 319.* National Sample Survey Organisation, Government of India.

Ghosh, J. K., Maiti, P., Rao, T. J., & Sinha, B. K. (1999). Evolution of statistics in India. *International Statistical Review, 67*(1), 13–34.

Ministry of Finance. (1952). *The national sample survey, General report No. 1 on the first round, October 1950–March 1951.* The Department of Economic Affairs, Government of India.

MoSPI. (2001). *Concepts and definitions used in NSS, Golden jubilee publication.* National Sample Survey Organisation, Government of India.

MoSPI. (2008). *Annual Report. The National Statistical Commission. 2007–2008.* The Government of India.

MoSPI. (2012). *Operational characteristics of unincorporated non-agricultural enterprises (excluding construction) in India, NSS 67th round, July 2010–June 2011, Report number 546.* National Sample Survey Office, Government of India.

MoSPI. (2013). *Economic characteristics of unincorporated non-agricultural enterprises (excluding construction) in India, NSS 67th round, July 2010–June 2011, Report number 549.* National Sample Survey Office, Government of India.

[32] Coverage of the Annual Survey of Industries (ASI) extends to the entire Factory Sector comprising industrial units (called factories) registered under the Sections 2(m)(i) and 2(m)(ii) of the Factories Act, 1948, wherein ten or more workers are working or were working on any day of the preceding twelve months with the aid of power or wherein twenty or more workers are working or were working on any day of the preceding twelve months without the aid of power. Enterprise surveys as per the NSS exclude the establishments within the coverage of ASI.

MoSPI. (2018). *Economic characteristics of unincorporated non-agricultural enterprises (excluding construction) in India, Report number 582, NSS 73rd round, July 2015–June 2016.* National Sample Survey Office, Government of India.

MoSPI. (2019). *Annual report, Periodic labour force survey (PLFS), July 2017–June 2018.* National Statistical Office, Government of India.

MoSPI. (2022). *Instructions to Field Staff. Volume-I. Household Consumption Expenditure Survey, July 2022–June 2023.* National Statistical Office, Government of India.

MoSPI. (2023). *Multiple indicator survey in India, NSS 78th round, 2020–2021, NSS report number 589.* National Sample Survey Office, Government of India.

MoSPI. (2024). *Annual report, Periodic labour force survey (PLFS), July 2023–June 2024.* National Sample Survey Office, Government of India.

Murthy, M. N. (1967). *Sampling theory and methods.* Statistical Publishing Society.

Murthy, M. N., & Roy, A. S. (1975). Development of the sample design of the Indian national sample survey during its first twenty-five rounds. *Sankhya, 37*(Series C, Pt. 1), 1–42.

Rao, T. J. (2012). *Origin and Activities of NSS, Government of India.*

Rao, T. J. (2016). MAHALANOBIS: Chief Architect of National Sample Survey and Related Contributions. *Sarvekshana*, 100th issue, pp. 35–50.

Chapter 3
Glimpses of Sample Design

3.1 General

The National Sample Survey Report Number 5 titled 'Technical Paper on Some Aspects of the Development of the Sample Design' by D.B. Lahiri—issued by the Department of Economic Affairs, Government of India (March, 1954)[1]—gives a broad account of the development of the sample design in the early period of the NSS since its inception during the year 1950. Subsequently, Murthy and Roy (1975) have discussed at length the development of the sample design of the NSS during the first twenty-five rounds i.e., till 1970–71. Thereafter, in February 2004, the MoSPI brought out a technical note titled 'Evolution of the Sample Design in NSS from 1st to 55th Round—A Critical Review'.[2] In 2025, MoSPI has brought out two Diamond Jubilee Publications titled 'Household Surveys of NSS: A Journey of 75 Years' and 'Enterprise Surveys of NSS: A Journey of 75 Years' (see MoSPI 2025a, 2025b) giving some details about the household and enterprise surveys conducted during the last 75 years with broad descriptions of sample designs adopted in NSS 55th round and onwards. However, this chapter gives a comprehensive overview on how the sample design in the NSS evolved over time since its inception till today considering quite a few NSS rounds—starting with some early rounds including the 1st round—and some important subjects of enquiry after dividing the 75-year survey period of the NSS into four phases.

Broadly speaking, the basic design in the NSS has remained the same i.e., a stratified two-stage/three-stage design with villages and urban blocks mostly constituting

[1] See The Department of Economic Affairs (1954).

[2] The note was prepared by K. Sankaranarayanan, ex-Joint Director of the NSSO. https://www.mospi.gov.in/sites/default/files/publication_reports/design_golden.pdf.

The original version of the chapter has been revised: New paragraph and reference have been updated. A correction to this chapter can be found at
https://doi.org/10.1007/978-981-96-7320-9_10

G C Manna, *75 Years of the Indian National Sample Survey*, India Studies in Business and Economics, https://doi.org/10.1007/978-981-96-7320-9_3

the sample FSUs and households/enterprises[3] (or plots of land in case of crop surveys) as the USUs. However, there have been changes from time to time in the methods of both stratification (with deeper stratification of FSUs over time) and sampling of FSUs with some illustrations given in the preceding chapter. As regards the sampling frame of FSUs, in case of rural areas, generally the lists of villages as per the latest population census have been used. With regard to urban areas the lists of census enumeration blocks mostly served as the frame for sampling of FSUs (i.e., urban blocks) till NSS 15th round and thereafter the Urban Frame Survey (UFS) lists of UFS blocks have been generally used since NSS 16th round. In case of enterprise surveys, after the Economic Census (EC) was launched in the year 1977, reliance has been on the use of lists of villages/urban blocks containing information on number of enterprises and workers by industry as per the latest Economic Census (EC) as the sampling frame of FSUs since NSS 33rd round (1978–79). To tackle different subjects of enquiry in a round, the practice has been to collect data for different subjects generally from different sets of households (and often with a different stratification of households for different subjects of enquiry) within the same sample FSUs.

A cursory look to the list of subjects covered so far in the NSS (Statement 2.1) will reveal that quite a few subjects were taken up in each round during the first twenty-five years of the NSS i.e., till NSS 29th round (July 1974–June 1975). Thereafter, the number of subjects of enquiry in a round has been generally much less. This chapter is primarily intended for two sets of readers namely (i) students and researchers/ statisticians, and (ii) the users of NSS data, who may be interested in knowing how the sample design evolved over time in the NSS. In this attempt to give a fair overview of the sample design on different subjects covered in the past 75 years, the chapter has unavoidably become a bit lengthy one. The period of 75 years of the NSS till today is divided into four phases as follows with the rationale behind such periodization:

- *the initial phase* from the year 1950 till middle of 1972 i.e., the year prior to the establishment of the NSSO[4];
- *the second phase*[5] from 1972 till 1999–2000 with the terminal year coinciding with the quinquennial survey period of NSS 55th round on employment–unemployment and household consumption expenditure [NSS 55th round had made

[3] In fact, the NSS considers establishments (i.e., the units located in single physical locations)—and not enterprises which consist of more than one establishment in case of multi-establishment enterprises—as the ultimate sampling units in its various surveys. However, to maintain consistency with the NSS terminology, the term 'enterprise' has been used in many places in this book for the convenience of the readers. In certain places, the term 'establishment' has also been used interchangeably.

[4] Although the NSSO was established as per the March 1970 Resolution of the Government of India, its full operationalization did not take place till 1972 i.e., prior to launching of the NSS 27th Round. The NSSO came into existence with the Governing Council in the helm of the affairs relating to the decisions on all aspects of the surveys including the finalization of the survey design. The first Chairman of the Governing Council was V.M. Dandekar. The first few rounds of this initial phase had explored various alternative stratification and sampling strategies.

[5] In the beginning of this second phase, the popular '*quinquennial surveys*' on employment–unemployment and household consumption expenditure as part of the overall NSS survey programme got introduced during the 27th round of the NSS (October 1972–September 1973).

certain significant changes in the reference period of data collection relating to household consumption expenditure which subsequently led to certain controversies (see the discussion on NSS 55th Round under the Second Phase in this chapter) with corrective measures taken in the next quinquennial survey on the topic];
- *the third phase* from 2000–01 till the year 2016–17; and
- *the fourth/last phase* since 2017–18 which is a period of some high frequency surveys like the Periodic Labour Force Survey (PLFS) launched from 2017–18 and Annual Survey of Unincorporated Sector Enterprises (ASUSE) introduced during 2021–22.

While selecting the round numbers for highlighting the sample design in each of the four phases, due consideration has been given to cover important subjects of enquiry, the rounds involving special features/changes in the survey methodology and other distinguishing features of the rounds/surveys.

3.2 The Initial Phase (1950 to 1972): NSS Round Numbers 1–26

- **The First Round of NSS (October 1950–March 1951)**

[Wages, land utilization, household enterprises, household assets and liabilities, consumer expenditure, prices.]

The NSS 1st round covered only rural areas of the country. Some of the major subjects covered in this round included land utilization, household enterprises, household assets and liabilities and consumer expenditure. The pre-requisite to any sampling exercise is the availability of a proper frame. Before sampling the geographical locations in the 1st round of NSS, it was found that the material available was different for different regions. Accordingly, different types of frames and different sampling methods were used in the 1st round for selection of sample villages in different parts of the country. A stratified two-stage design was adopted with villages as the FSUs and households as the second-stage units for the socio-economic surveys.

On the lack of material to develop a sampling frame for selection of villages, the NSS General Report Number 1 on the 1st round mentions that 'attempts were first made to collect large scale detailed maps showing the village boundaries, but it was found that such detailed maps were not available for a large part of the country. Attention was then turned to collect lists of villages with individual figures of population and area. This was also not possible for the whole of India. Due to the partition of India in 1947 and the merging and reorganisation of the States during 1948, the position had become complicated and for certain districts or tehsils even the figures given in the 1941 Census Reports could not be used'.

All the States of the country excluding Jammu & Kashmir and Sikkim were first divided into 160 strata with each stratum formed 'on the basis of geographical contiguity and topographical homogeneity'. In the strata, 32 in number, where village-wise population figures were available, 4 sub-strata were formed, namely, villages with population of: 1–499 (sub-stratum 1), 500–999 (sub-stratum 2), 1000–1999 (sub-stratum 3), and 2000 or more (sub-stratum 4), which resulted in 128 (i.e., 32 × 4) ultimate strata. Thus, altogether, 256 strata were formed for sampling of villages as the first-stage units.

The sample of the 1st round was composed of two groups of villages where two different sets of schedules were canvassed. One set i.e., the one prepared by the ISI was used for canvassing in the first group of 1,189 villages and the other set of schedules prepared by the Gokhale Institute of Politics and Economics was canvassed in the second group of 644 villages.[6] The final allocation of sample villages to different strata was in proportion to total rural population of the respective strata as per 1941 census. The sample allocation for each stratum was in multiples of 3 so that the allocation of sample villages between the two sets of schedules could be in the ratio of 2:1 approximately. The method of sampling of villages was different for different strata depending upon the nature of information that was available. In fact, three different sampling methods were used namely equal probability, probability proportional to population and probability proportional to area.

As regards the sampling of households, all the households in the sample village were listed out of which a random sample of 80 households was selected randomly for collecting information relating to occupation. These 80 households were divided into two groups i.e., agricultural and non-agricultural. From each group a sample of 8 households was selected randomly to canvass the household schedule for general particulars. The schedule on household enterprises was canvassed in 2 of the 8 agricultural households and 3 of the 8 non-agricultural households selected at random. Out of 6 remaining agricultural households, 1 was selected at random and from the remaining 5 non-agricultural households, 2 households were selected at random for collecting information on consumer expenditure.

The land utilisation survey was carried out in the same sample of villages with clusters of plots as the second-stage units. For this purpose, 'on the basis of village maps showing plot boundaries, 20 clusters of 5 plots each were selected with probability proportional to area of cluster'.[7]

- **NSS 7th Round (October 1953–March 1954)**

[Village statistics, land utilization, agriculture and animal husbandry, non-household manufacturing establishments, small-scale manufacture and handicrafts, transport, trade, profession service and financial operations, consumer expenditure, prices, opinion on out-turn of rice in village.]

As mentioned in the previous chapter (see Sampling of FSUs), instead of a two-stage design that has been generally followed in the NSS with villages/urban blocks as the FSUs, NSS 7th round adopted a three-stage design with teshils in rural areas and towns in urban areas forming the FSUs while villages/urban blocks constituting

[6] For details including the contents of two sets of schedules, see Ministry of Finance (1952).

[7] For further details, see Ministry of Finance (1952).

the SSUs.[8] Further, NSS 7th round made certain improvements in the sampling of households to net adequate number of enterprises in the transport sector.[9] Under these backgrounds, it may be useful to see the broad sample design adopted in the seventh round.

NSS 7th round covered the whole country except Jammu and Kashmir, and Andaman and Nicobar Islands. To collect data on different subjects of enquiry, more than 40,000 households were visited of which 4,394 households were contacted to collect data on transport enterprises.

As regards the stratification in the rural areas, the country was first divided into 'natural divisions' as defined by the Registrar General, Ministry of Home Affairs, Government of India. The total all-India sample size of number of tehsils was allocated to different natural divisions in proportion to respective total consumer expenditure estimated based on the NSS 1st round data. The tehsils within each natural division were further divided into a few strata based on population density with the condition that total consumer expenditure of each stratum was roughly equal. Altogether, 240 strata were formed with each stratum having approximately same total consumer expenditure and a sample of 2 tehsils was allotted to each stratum. Sample tehsils were selected with probability proportional to area or population and with replacement. Similarly, from each sample tehsil, 2 villages were selected with probability proportional to area or population and with replacement. This resulted in a total sample size of 960 villages.

In urban India, each of the four big cities of Bombay, Calcutta, Delhi and Madras formed a separate stratum. The other cities and towns were grouped into 4 strata based on their population: 100,000 or more (stratum 1); 50,000–100,000 (stratum 2); 15,000–50,000 (stratum 3); and towns with population below 15,000 (stratum 4) separately for each of the four regions built up from six population zones adopted by the Registrar General. A total of 53 towns were selected with an allocation of 14, 9, 14 and 16 towns from stratum 1, stratum 2, stratum 3 and stratum 4 respectively. Total number of sample blocks was 444 which was allocated to strata of 4 big cities and of other cities and towns as per their population. From each stratum, the required number of sample blocks was selected with probability proportion al to total number of households in the blocks and with replacement.

The method of sampling of households was same for rural and urban areas. For the purpose of collection of data on household enterprises, the listed households were stratified according to broad industrial grouping by the productive enterprise of the household and a pre-specified number of households was selected for detailed

---

[8] A similar design was also adopted in NSS rounds 4 to 6 in rural areas and NSS rounds 3 to 6 and 8 in urban areas.

[9] As The Cabinet Secretariat (1958) mentions, 'In the four previous rounds, viz., third, fourth, fifth and sixth, the sample households for survey of different kinds of enterprises were selected from a common list of households in the sample villages......Since the proportion of households engaged in transport operations was very small, the number included in the sample was also very small.....In the seventh round, to obviate this difficulty, separate lists of households were prepared for different occupations and a large number of samples were selected from households which were offering transport services'.

enquiry. For example, for the data on transport enterprises, a sample of 10 households per village/block was selected from among the households owning transport enterprises.

- **NSS 9th Round (May 1955–November, 1955)**

[Village statistics, employment and unemployment, land utilisation, small scale manufacture and handicrafts, transport, trade, consumer expenditure, income and expenditure, prices.]

This was, in fact, the first NSS round which collected data on employment and unemployment situation in the country. The other subjects were land utilization, small scale manufacture and handicrafts, transport, trade, consumer expenditure, income and expenditure, and prices. The fieldwork of NSS 9th round was spread over a duration of six months from the middle of May 1955 to mid-November 1955. The survey covered the whole country except Sikkim and Andaman and Nicobar Islands.

As regards the stratification, a little less than 300 strata were formed in rural India with one district or group of contiguous districts forming a rural stratum. Altogether 1,624 villages were selected as the FSUs after allocating them to different strata in proportion to respective rural populations and the stratum level sample size was adjusted to multiples of 4 to select the sample villages in the form of 4 independent sub-samples out of which the first two sub-samples were surveyed during the first three months of the survey period and the remaining two sub-samples were covered during the next three months of the survey. The sample villages from each stratum were selected by probability proportional to population (1951 Census) with replacement. From each sample village, a sample of 10 households was selected by systematic sampling from the list of households prepared by the field investigators in respective sample villages.

The urban areas of the country comprising 3,018 towns were divided into 94 strata as against only 20 strata formed in eighth round. Each city with a population of 3 lakhs or above and also each capital town of 'part A and part B States excluding Shillong' formed a separate stratum by itself.[10] The other set of strata comprised the remaining towns in each of the States. Some departure was made in respect of greater Calcutta (excluding Calcutta city and Howrah town) and greater Bombay (excluding Bombay city). While greater Calcutta was sub-divided into seven strata, greater Bombay formed a separate stratum. In the urban areas, the 1951 census blocks comprised the FSUs. The all-India urban sample size of 2,108 blocks was allocated to different States in proportion to non-agricultural population which was further allocated over different strata within the State in proportion to non-agricultural population with the stratum level allocations rounded off to multiples of 4 as in the case of rural areas to facilitate selection of blocks in the form of 4 independent sub-samples of which the first two sub-samples were covered during the first three months of the survey period. From each stratum, the sample blocks were selected by circular systematic sampling with equal probability. The listed households in each census block were grouped into 2 strata: those comprising households having at least one unemployed

[10] For further details, see The Cabinet Secretariat (1959).

person (first stratum), and the remaining households (second stratum). A sample of 10 households from the first stratum and 6 households from the second stratum was selected by systematic sampling from each block.

- **NSS 14th Round (July 1958–June 1959)**

[Village statistics, population, births and deaths, employment, land utilization and yield surveys, small-scale manufacture and handicrafts, income and expenditure, prices.]

NSS 14th round is the first round which was spread over one year as against all previous rounds confining to a few months only. In this round, apart from village statistics, the other subjects of enquiry included population, births and deaths, employment, land utilisation and yield, small scale manufacture and handicrafts, income and expenditure, and prices. Different schedules of enquiry were prepared for the different subjects. Most of these subjects were covered in some of the previous NSS rounds as well.

The survey covered the entire country except Andaman and Nicobar Islands, Amindive and Laccadive Islands, the North-East Frontier Agency, and the rural areas of Ladakh district of Jammu and Kashmir. In this round, the States of 'Assam, Bihar, Bombay, Kerala, Orissa, Punjab and Uttar Pradesh participated in the work of the NSS on full matching basis and the state of Andhra Pradesh participated on half-matching basis.' The central sample consisted of 2,616 villages and 2,228 urban blocks. The survey in urban areas was confined only to the subjects of Income and expenditure; Small-scale manufacture and handicrafts; and Employment and unemployment. To take care of seasonal fluctuations, particularly with regard to subjects like employment–unemployment, household enterprises, income and expenditure, and crop statistics, the survey period was divided into six sub-rounds of two months' duration each and all the sample villages were visited in each of the six sub-rounds.

As regards the rural stratification, each State was stratified by grouping contiguous tehsils which were homogeneous with respect to population density, altitude above sea level and cultivation of food crops with the condition that each stratum had approximately equal population. The number of strata formed within a State is one-twelfth of the sample allocation to the State since 2 investigators were posted in every stratum with each one of them covering a sub-sample of 6 villages. From each stratum, the sample villages were selected by circular systematic sampling with equal probability. Large villages were divided into a few hamlet-groups out of which one was selected at random to prepare the frame of households for different enquiries.

The data collection/fieldwork for different subjects of enquiry was organized as follows: 'collection of data on population, births and deaths from all the sample villages in the first and the last sub-rounds, on prices in one-third of the sample villages in each sub-round, on village statistics in all the villages in any sub-round, and on land utilisation in all the sample villages and yield survey in one-third of the sample villages in each crop season'. For employment–unemployment enquiry, a sample of 24 households per village was selected by circular systematic sampling after arranging the households according to its size i.e., 1 to 4 and 5 or above. The employment data were collected from the same set of sample households once in every two months. But the same set of households could not be visited for the data on income and expenditure and household enterprises.[11]

The stratification for urban areas was exactly the same as in 9th round with 1951 census blocks comprising the FSUs. Further, all the sample blocks of the 9th round were revisited in this round except for the sample blocks of Madras city and Jammu and Kashmir for which fresh sampling frames were available. From each sample block, a random sample of households was selected for three different subjects of enquiry from the frame of households as per the listing exercise. As regards collection of data on employment–unemployment, all the households in the sample block were arranged with those having unemployed members coming first followed by other households. Thereafter, a linear systematic sample was drawn using the random start and sampling interval provided in the list of sample blocks.

- **NSS 17th Round (July 1961–June 1962)**

[Population, births and deaths, family planning, morbidity, employment, urban labour force, ownership of land and operational holdings, land utilization, yield surveys, capital formation, consumer expenditure, prices.]

The land utilization survey and crop cutting experiments (for measuring the yield rate) formed two important components of the NSS 17th round in order to have estimates of production of seven major cereal crops namely rice, wheat, jowar, bajra, ragi, maize and barley. While data on land utilization were collected by 'actual observation of the utilisation of the selected plots', the same on yield of crops were collected by 'harvesting a circular area of radius 4′ in selected plots' [Source: NSS Report Number 106, Seventeenth Round]. The survey covered the whole country excluding a few areas.[12]

The land utilization survey adopted a stratified two-stage design[13] with villages as the first-stage units and clusters of plots as the second-stage units. However, the crop cutting experiments adopted a multi-stage design with different stages of sampling units being the villages, clusters of plots, sub-plots growing the specified crops and circular area of radius 4′ respectively. For the selection of the ultimate sampling unit i.e., the circle of radius 4′, the centre of the circle was selected at random. The driage experiments were carried out to arrive at the driage factor for obtaining the final estimate of production of dry crop. For this purpose, as NSS Report Number 106

---

[11] See The Cabinet Secretariat (1962) for further details regarding sampling of households for canvassing the schedules on income and expenditure and household enterprises.

[12] The excluded areas were Andaman & Nicobar Islands, Laccadive, Minicoy and Amindivi Islands, Ladakh district of the then State of Jammu & Kashmir and certain areas of Manipur and North-East Frontier Agency.

[13] Altogether 216 strata were formed for sampling of villages with 48 regions being the basic strata where each region comprised a group of contiguous districts within a State having similar crop pattern and population density. Each such region was further divided to form compact strata by grouping contiguous tehsils having similar population density such that each ultimate stratum had roughly equal population. The overall sample size of 3,888 villages was allocated over strata based on the joint consideration of rural population (1951 Census), area under food crops (1954–55) and the available strength of investigating staff.

mentions: 'The harvesting of crops in the sample cut of radius 4′ was done in two stages—first in a circle of radius 2′ 3″ and then in the annular portion. The yield of crop from the circle of radius 2′ 3″ from each sample cut was used for driage experiments…'.

- **NSS 19th Round (July 1964–June 1965)**

[Village statistics, population, births and deaths, housing condition, urban labour force, employment, unemployment and indebtedness of rural labour households, land utilisation, yield survey, integrated household schedule-detailed and abridged, prices.]

NSS 19th round is an important round in the sense that apart from other subjects of enquiry (like population, births and deaths; employment; small scale manufacture and handicrafts; and income and expenditure), this round also covered land utilization and crop yield. Although land utilization survey (LUS) was carried out as part of the 1st round itself, the crop yield survey was introduced only in the 10th round of the NSS (December 1955–May 1956) for obtaining the yield rates of some major food grains. Both these subjects were covered till 24th round (July 1969–June 1970) barring one or two NSS rounds. Another distinguishing feature of this round was that it canvassed an Integrated Household Schedule for eliciting an overall picture of the socio-economic activities of the sample households (through Schedule No. 16) apart from some other schedules namely Schedule 5.0 (Land utilization survey) in rural areas for the measurement of land area by crops, Schedule 5.1 (Crop cutting experiments) in rural areas for estimation of yield rates, Schedule 10 (Urban labour force), Schedule 10.1 (Employment, unemployment and indebtedness of rural labour households), and Schedule 12 (Population, births and deaths) canvassed in both rural and urban areas.

As regards the broad sample design, it was a stratified two-stage design for the household enquiries with villages/urban blocks as the FSUs and households as the SSUs. The sample design was self-weighting for the household enquiries at the state level for rural areas and at the stratum level for the urban areas.

The all-India sample size of FSUs was fixed at 8,472 villages 4,572 blocks in the central sample. The Indian economy being highly dependent on agriculture which is seasonal fluctuations, the entire survey period was divided into 6 sub-rounds with each having a duration of 2 months and equal number of FSUs was covered in each sub-round. All the field investigators were divided into 2 parties with party 1 covering the FSUs belonging to sub-samples 1 and 3 while party 2 investigators covered the FSUs of sub-samples 2 and 4. This scheme helped in studying the components of total margin of error, namely, variations between sub-samples, between parties and within parties.

The total sample size of 8,472 villages was allocated to States/UTs by jointly considering the rural population, area under food crops and investigator strength of respective States/UTs with a minimum sample allocation of 360 villages to most of the States. Rural strata of compact geographical areas within 'agricultural regions' of a State fulfilling certain other conditions (like the parts of the stratum having similar population density, altitude and transport facilities) were formed such that each stratum had roughly equal size (the size being linked to the FSU size related to population). The total number of rural strata within a State was arrived at by dividing the total sample size of villages for the State by 24 (rounded off to an integer) and from each stratum 24 villages were selected by circular systematic sampling with PPS in the form of 4 independent sub-samples of 6 villages each.

Urban strata were formed within each State/UT by grouping towns according to population as per 1961 census. Generally speaking, 2 strata were formed within a State/UT, namely stratum 1 comprising towns with population 50,000 or more and stratum 2 comprising the remaining towns. However, in the State of Maharashtra, one additional stratum of Bombay city was formed by separating out this city from stratum 1. The all-India sample size of 4,572 blocks was allocated to different States x strata by considering their relative share in 1961 census urban population with double weight to stratum 1 towns. For most of the towns, the UFS blocks served as the sampling frame of urban blocks. However, for 700 towns out of total number of 2,703 towns as per 1961 census, the UFS frame was not ready and accordingly the 1961 census lists of blocks were as the sampling frame for these towns.

The sample FSUs i.e., villages/blocks were selected circular systematically with PPS, the measure of size being related to the population. The SSUs from each FSU (or one randomly selected hamlet-group/sub-block for large FSUs) were selected linear systematically with equal probability. For this purpose, the listed households were arranged according to 6 household (HH) classes. These 6 classes in the villages were formed as explained below Boxes 3.1R and 3.1U by jointly considering the household size and means of livelihood (ML) of the households in rural areas and the household size, number of unemployed persons and number of self-employed persons in urban areas:

The above arrangement ensured representation of households having small and large sizes and different means of livelihood (rural)/number of unemployed or self-employed members (in case of urban) proportionately in the sample.

Out of 22 sample households selected linear systematically from a village, Schedule 16 i.e., the Integrated Household Schedule was canvassed in a sub-sample of 2 households on an average per village and in all the remaining sample households, schedule 12 (Population, births and deaths) was canvassed. Schedule 10.1 was canvassed in all the rural labour households among those selected for schedule 12.

Box 3.1R Method of arrangement of listed households before sampling for rural areas, NSS nineteenth round

HH class	1	2	3	4	5	6
No. of members	1–4	1–4	1–4	5 or more	5 or more	5 or more
ML Code	3	2	1	1	2	3

Note ML Code 1: Agricultural labour; ML Code 2: Self-employed in non-agriculture; ML Code 3: Others.
Source The Cabinet Secretariat (1968)

Box 3.1U Method of arrangement of listed households before sampling for urban areas, NSS nineteenth round

HH class	1	2	3	4	5	6
Number unemployed	1 or more	1 or more	0	0	0	0
No. of members	1–3	4 or more	4 or more	4 or more	1–3	1–3
Number self-employed	*	*	0	1 or more	1 or more	0

Note *HH classes 1 and 2 may or may not contain any self-employed persons
Source The Cabinet Secretariat (1968)

The procedure of sampling of households for urban blocks was very much similar to that adopted for the villages.

As regards the crop survey, sample design was stratified two-stage for the subject of land utilization i.e., crop area survey. However, for crop cutting experiments i.e., the survey on yield rates, a stratified four-stage design was adopted with the sampling units at various stages being villages, clusters of plots, crop plots and circular cuts respectively. The crop surveys and household enquiries were conducted in the same set of villages. The LUS was conducted in all the 24 villages of a stratum. On the contrary, the crop yield survey (i.e., crop cutting experiment) was carried out in only 6 villages per stratum i.e., 3 out of 6 sample villages of sub-samples 1 and 2 only. The price enquiry was conducted in the fixed set of 419 sample/price villages which were independent of the general sample villages of this round comprising 8,472 villages.

3.3 The Second Phase (1972 to 1999–2000): NSS Round Numbers 27–55

- **NSS 27th Round (October 1972–September 1973)**

[Seasonal migration, number of pucca, semi-pucca and kutcha houses, employment and unemployment, current building activity in rural areas, consumer expenditure, prices.]

NSS 27th round was the first 'quinquennial survey' covering the subjects of employment–unemployment and household consumer expenditure. The subsequent rounds which covered these two subjects as a part of the quinquennial series are NSS round numbers 32, 38, 43, 50, 55, 61, 66 and 68. In fact, NSS 68th round (2011–12) happened to be the last round in the quinquennial series. NSS 27th round also covered information on seasonal migration, number of pucca, semi-pucca and kutcha houses, and current building activity in rural areas.

All the States excluding Nagaland and West Bengal and some UTs participated in this round at least on equal matching sample basis to have improved State/UT level estimates based on central and state samples. The survey covered the whole country except 'Ladakh district of Jammu & Kashmir state, rural areas of Pal and Samuri tehsils of district Surguja and Prijapur, Dantwara, Konta and Narayanpur tehsils of district Chanda and Welghat tehsil of district Amravati of Maharashtra; rural areas of Nagaland; disturbed villages of Tripura, rural areas of the Union Territories of Arunachal Pradesh; Andaman and Nicobar Islands, Mizoram, Laccadive, Minicoy and Amindivi Islands, and Dadra & Nagar Haveli'. In the central sample, altogether 9,088 villages and 4,832 urban blocks were selected as the FSUs. The design was made 'self-weighting' at the sub-stratum and region level for rural areas and sub-stratum and state level for urban areas.[14]

Coming to the sample design for the rural areas, first of all the list of villages as per 1971 Population Census served as the sampling frame of FSUs (i.e., villages). In rural India, 392 strata were formed. For this purpose, the country was first divided into 66 agricultural regions by grouping districts having similar population density and crop pattern within each State/UT. Thereafter, strata were formed within each region by grouping contiguous tehsils having similar population density, crop patter, altitude and also good transport and communication facilities amongst them. Strata were formed in such a fashion that total population of each stratum became almost the same within a region.

The all-India sample size of 9,088 villages was allocated to the States by jointly considering rural population, area under cereal crops and available investigator strength with a minimum allocation of 360 villages excepting some smaller States and UTs. Within a State/UT, number of strata formed was 'n' divided by 24 where 'n' denotes the sample size of number of villages allotted to a stratum, with a sample allocation of 24 villages to each stratum drawn in the form of 2 independent sub-samples of 12 villages each generally. The sample villages from a stratum were selected by circular systematic sampling with equal probability after arranging the tehsils in a serpentine order.[15] The large villages were divided into 2 or more HGs by roughly equalizing population and one HG was selected at random for listing purposes. For canvassing the schedules on employment–unemployment and consumer expenditure, all the listed households were grouped into 3 sub-strata as per their means of livelihood namely self-employed in non-agriculture, rural labour and others with an allocation of 2, 5 and 5 households respectively to the three sub-strata. Sampling of households from each stratum was done linear systematically using the interval and random start specified for the purpose to make the design self-weighting at the region level for each sub-stratum. Both the schedules on employment–unemployment and consumer expenditure were canvassed in the same sample households.

For urban areas, the sampling frame for selection of urban blocks as the FSUs comprised the lists of UFS blocks along with the lists of 1971 census blocks (1961 census blocks in respect of Tamil Nadu and West Bengal), for those towns for which UFS frame was not available. Two strata were formed within each State/UT namely towns with population (1971 census) less than 50,000 forming stratum 1 and other towns/cities forming stratum 2. However, each of the 4 big cities namely Bombay,

[14] In a self-weighting design, the weights i.e., multipliers for all the sample households become the same for estimating the population totals, which makes the calculation quite simpler. For quite a few rounds since the fourth round 'the NSS design has been made self-weighting for one or more household enquiries of each round'. Murthy and Roy (1975) discuss in detail the procedure of self-weighting at the field stage and at the tabulation stage.

[15] For further details of sample design, see Department of Statistics (1974).

Calcutta, Delhi and Madras formed a separate stratum. This procedure resulted in 50 urban strata for the country as a whole.

The all-India sample size of 4,832 blocks was allocated to States in proportion to 1961 census urban population with a minimum allocation of 144 blocks to each State except some smaller States and UTs. State level sample allocation was further allocated among different strata in proportion to population. Stratum level allocations were in multiples of 8 to facilitate sampling of blocks in the form of 2 independent sub-samples of equal size coupled with 4 sub-rounds. The required number of sample blocks from each stratum was selected by circular systematic sampling with probability proportional to size, the size being population of the block rounded off to the next multiple of 1000, after arranging the towns by region and blocks within the towns as per geographical contiguity. In the large blocks, sub-block selection was resorted to as was the case for large villages. Finally, the listed households were grouped into 2 sub-strata: self-employed (sub-stratum 1) and others (sub-stratum 2). The households of sub-stratum 1 were arranged into 2 classes: self-employed in manufacture and other self-employed households. The total sample size of 12 households per block was allotted to each sub-stratum in equal number (i.e., 6 households) and the sample households were selected by linear systematic sampling using the interval and random start made available in the sample list of selected blocks. The sample design was made self-weighting at the State and sub-stratum level. As in rural areas, both the schedules were canvassed in the same sample households.

- **NSS 29th Round (July 1974–June 1975)**

[Employment, unemployment and indebtedness of rural labour households, small scale manufacture and handicrafts, mining and quarrying, trade, hotels and restaurants, transport, service, construction, consumer expenditure, income, availability and extent of utilization of electricity in rural areas, prices.]

As mentioned earlier, NSS 29th round was presumably the last round covering many subjects of enquiry including household enquiry on self-employment in various types of non-agricultural enterprises, employment–unemployment and indebtedness of rural labour households, consumer expenditure, income, availability and extent of utilization of electricity in rural areas apart from prices. The survey adopted a two-stage design with villages/census enumeration blocks as the FSUs and households as the SSUs (second-stage units). The sample design was made 'self-weighting' for all the schedules of enquiry at region and State level separately for rural and urban areas. As part of the central sample, 8,512 villages and 4,872 blocks were selected.

In rural areas, 1971 census list of villages was used as the sampling frame of FSUs. The whole country was divided into 524 basic strata such that they did not cut across district boundaries. With certain exceptions, generally each district having a rural population (1971 census) of less than 1.5 million formed one basic stratum. The districts with population of 1.5 million or more were divided into two or more basic strata by grouping contiguous tehsils and homogeneous as far as possible with respect to rural population density and crop pattern such that each such basic stratum had a population of less than 1.5 million. The sample allocation of FSUs over basic strata within each State/UT was in proportion to rural population with stratum level allocations as multiples of 4. The sample villages from each basic stratum were selected with probability proportional to population (1971 census) and with replacement.

In the urban areas, 1971 census enumeration blocks formed the FSUs. Rural basic stratum was considered as basic stratum for urban also. However, some such basic strata/districts with low urban population were merged with neighbouring districts within the same region to form the urban basic strata with sizeable urban population. In this manner 386 urban strata were formed. Each city with 1971 census population of 100,000 or more formed a separate stratum and all the towns of a basic stratum were grouped into 3 sub-strata as per their population size, namely less than 20,000 (sub-stratum 1); 20,000 to 49,999 (sub-stratum 2); and 50,000 to 99,999 (sub-stratum 3). Sample allocations for different ultimate strata were as per their proportionate shares in urban population. The sample blocks were selected with probability proportional to population (1971 census) and with replacement.

The procedure of selection of households for enterprise schedules was same for both rural and urban areas. During the listing exercise a household having at least one self-employed member in any of the five non-agricultural household enterprises[16] under the survey coverage during the preceding 365 days was considered as an 'enterprise household'. Sample households for enterprise enquiries (with 5 different schedules of enquiry design for data collection) were selected linear systematically from the list of enterprise households. In each sample household all the relevant enterprise schedules (for which the household was eligible) were canvassed.

- **NSS 30th Round (July 1975–June 1976)**

[Village statistics, livestock number and products, consumption of livestock products, livestock enterprises, prices, survey on railway travel.]

Realizing the need to fill up the data gaps in animal husbandry statistics, the Governing Council of NSSO decided to collect data on livestock number and products and livestock enterprises in NSS 30th round. The sample size for central sample was 8,512 villages and 9,744 urban blocks. It is important to note that the urban sample size of FSUs in this round was one among the highest sample sizes adopted so far in the history of NSS. The sample size during NSS 18th to 29th rounds was in the range of about 4,500 to 5,000 urban blocks.

To collect data on number of livestock/poultry and their products, three separate schedules viz. Schedules 19.1, 19.2 and 19.3 were respectively used for bovines & poultry, pack animals (namely horse, donkey, mule, camel & elephant), and ovines & pigs. Schedule 19.4 was used to collect data on consumption of livestock products by the households. For production and cost aspects of the livestock enterprises, Schedule 20 was canvassed. However, enterprises registered under Sects. 2 m(i) and 2 m(ii) of the Factories Act, 1948 were out of coverage. For enterprises like dairy, poultry, sheep and goat raring, piggery, cattle and buffalo breeding, the scale of

[16] Enterprises were classified into 5 types i.e., (i) small-scale manufacturing and repair services; (ii) transport, storage and communication; (iii) trade, hotel and restaurant; (iv) services and construction and (v) mining and quarrying.

operation was taken into consideration to exclude tiny household farms which did not possess distinguishing features of an enterprise. To ensure adequate number of households possessing different type of livestock and poultry, a sample of as large as 24 households per village was considered. The design was self-weighting in urban areas with about 24 households on an average per sample block. In the non-household sector, three types of livestock enterprises i.e., dairy, poultry and piggery operated by institutional establishments were considered and Schedule 20.1 was designed to collect the relevant information. As regards the migratory flocks of animals, the survey had provision to collect information on number of animals possessed, receipt from important products and other details from the person(s) accompanying the flock by locating such flocks in the sample village/block or in its vicinity.

The sample design was a stratified two-stage one with villages/urban blocks as the FSUs and households as the SSUs for Schedule 19 series and for Schedule 20. However, for Schedule 20.1, SSUs were non-household establishments.

The entire survey period of one year was divided into 4 sub-rounds of three months each and equal number of sample villages/blocks was allotted to each sub-round for the purpose of data collection.

For the purpose of stratification, the entire country was divided into a number of 'basic strata' so that they did not cut across district boundaries. Generally speaking, each district with less than 1.5 million rural population as per 1971 census formed a separate basic stratum. However, the larger districts in terms of population were divided into 2 or more basic strata by grouping contiguous tehsils, homogeneous as far as possible in respect of density of rural population and crop pattern, such that each basic stratum had a rural population of less than 1.5 million. A similar procedure was followed for the urban stratification also. However, some of the basic strata having very low urban population were merged to form urban strata with sizeable urban population. Out of 374 districts in the country at that time, 366 districts were under the survey coverage and the total numbers of rural and urban strata were 520 and 386 respectively.[17]

The all-India sample size of villages/urban blocks was allotted to respective States/UTs considering their respective investigator strengths and rural/urban population. This sample size within a State/UT was further allocated to various basic strata as per their share in total population. Sample size for each basic stratum was in multiples of 4 to allocate equal number of sample FSUs to each sub-round.

For rural areas, 1971 list of villages served as the sampling frame of FSUs. For the urban areas, it was a mixture of UFS list of blocks and the list of 1971 census blocks depending upon the availability of the UFS frame. Both villages and urban blocks were selected with pps (the size being population of the village/block as per the sampling frame) and in the form of 2 independent sub-samples. As in the other NSS rounds, hamlet-group/sub-block formation was resorted to in the large villages/blocks and one of them was selected at random for preparing the frame of households/establishments.

[17] Department of Statistics (1978a)

For canvassing Schedules 19.1 to 19.4, all the listed households in the village/block were arranged with the households possessing any livestock or poultry appearing first followed by other households. From this arranged frame, a sample of 24 households from each village was selected by circular systematic sampling. In each urban block, a sample of households (24 on an average) was selected linear systematically with a pre-specified sampling interval. All the selected households were surveyed for these four schedules.

Schedule 20 was canvassed in all the households usually engaged in the livestock and poultry enterprises with exclusion of certain types of enterprises as already mentioned. Schedule 20.1 was canvassed in all the non-household and non-registered establishments engaged in any of the activities of dairy, poultry and piggery.

- **NSS 34th Round (July 1979–June 1980)**

[Education, medical and health, unorganized sectors of trade, transport, hotels and restaurants, storage, warehousing and services, prices.]

The first Economic Census (EC) was conducted in the year 1977. A major objective to launch the EC was to provide the FSU (village/block) level information on number of establishments and workers by industrial category such that this database could be used as the sampling frame of FSUs in the enterprise surveys. NSS 33rd round was the first round when such FSU level database as per the first EC was used as the sampling frame of FSUs. While NSS 33rd round was devoted to the survey of unorganized manufacturing enterprises, the coverage of NSS 34th enterprise survey was much wider and accordingly it will be of interest to look at the sample design of this round and particularly the method of sampling of enterprises.

NSS 34th round covered small scale enterprises engaged in trade, hotels and restaurants, transport, storage & warehousing, and other services sector. It is important to mention that in the first EC enterprises were classified into two groups namely (i) own account enterprises (OAEs) i.e., enterprises without any hired worker and (ii) 'establishments' namely those operating with the help of at least one hired worker on a fairly regular basis. Establishments under category (ii) were further divided into two groups i.e., non-directory establishments (NDEs) (with total number of workers including household worker as 5 or less, and annual output/turnover/receipt less than Rs. 100,000) and directory establishments (DEs) with total number of workers including household worker as 6 or more and/or annual output/turnover/receipt of Rs. 100,000 or more. NSS 34th round was confined to only OAEs and NDEs (following the same definitions adopted in the EC) in the selected industries as specified above while the DEs were separately covered by the Central Statistical Organisation. Enterprises which operated for at least 30 days (15 days in case of seasonal enterprises) during the last 365 days and which existed on the date of survey were eligible for data collection. Government and public sector enterprises were outside the survey coverage.

The survey covered the entire geographical area of the Indian union excluding Sikkim, Andaman & Nicobar Islands, Dadra & Nagar Haveli, Lakshadweep, rural areas of Nagaland, and Ladakh district of Jammu & Kashmir. Further, two tehsils of district Surguja and four tehsils of district Bastar of Madhya Pradesh, three tehsils of district Chandrapur and one tehsil of district Amravati of Maharashtra were not covered. All the States and UTs participated in the survey at least on an equal matching sample basis.

A stratified two-stage design with census enumeration blocks (CEBs) as the FSUs and OAEs/NDEs as the SSUs (second-stage units) was adopted for both rural and urban areas. Entire country, excluding the areas left out of the survey coverage, was divided into 73 regions. These regions were formed within a State/UT, except in Gujarat, by grouping contiguous districts of similar agricultural profile. However, in Gujarat, regions were formed by grouping continuous tehsils considering the dry areas as well as the concentration of tribal population in the State. Each district formed a separate stratum except in Gujarat where parts of a district belonging to different regions were treated as separate strata.

The list EBs of 1977 EC was used as the sampling frame for selection of EBs as the FSUs. However, owing to non-availability of the EC frame, the list of UFS blocks was used as the sampling frame for urban areas of Nagaland and 1971 census list of villages was adopted as the sampling frame for rural areas of Chandigarh.

Each stratum was divided into 2 sub-strata with the following composition: EBs with at least one DE/NDE in any of the four industries namely trade, hotel & restaurant, transport, and storage & warehousing (sub-stratum 1); and the remaining EBs (sub-stratum 2). However, in rural areas of Gujarat, 3 sub-strata were formed by splitting sub-stratum 2 into two parts namely (i) those EBs having at least one DE/NDE in any remaining industry other than the four industries specified above (sub-stratum 2), and (ii) the residual EBs (sub-stratum 3).

The allocation of total number of sample EBs to a State/UT was broadly guided by the number of investigators available for the survey and the workload in terms of number of FSUs each investigator could cover in a year. In the central sample, altogether 6,661 EBs in the rural areas and 10,662 EBs in the urban areas could be surveyed. Total State/UT level sample size of EBs was allocated to rural and urban areas of the districts in proportion to total number of Des and NDEs in the four industries. Thereafter, district-wise sample size was allocated to sub-stratum 1 and sub-stratum 2 in the ratio of 5:1 separately for rural and urban areas. Sample EBs were selected from sub-stratum 1 by PPS (probability proportional to size), size being total number of DEs and NDEs in the four industries specified above as per the sampling frame/EC, and with replacement. From sub-stratum 2, sample EBs were selected at random (with equal probability).

Normally, entire sample EB was considered for survey i.e., listing of enterprises. However, the EBs with a population exceeding 1,000 or total number of DEs and NDEs exceeding 200 was divided into a few sub-blocks of nearly equal population or equal DE plus NDE content and one sub-block was selected at random for listing/preparation of frame of sample enterprises.

As regards the sampling of enterprises, altogether 11 schedules of enquiry were prepared for the purpose of canvassing in different types of enterprises in this round (see Box 3.2).

In each sample EB, all the OAEs eligible for schedule 2.51B were surveyed. In addition, a maximum of 12 NDEs were surveyed. Also, from each EB, 3 OAEs other

Box 3.2 Schedules canvassed in NSS thirty-fourth round

Sl. no	Schedule no.	Enterprise type
1	2.31B	Mechanised passenger & goods transport—NDE & OAE
2	2.32B	Non-mechanised passenger & goods transport—NDE & OAE
3	2.33B	Services incidental to transport—NDE & OAE
4	2.34B	Storage & warehousing—NDE & OAE
5	2.41B	Trade—NDE & OAE
6	2.42 B	Hotel & restaurant—NDE & OAE
7	2.51B	OAE in [other] service sector
8	2.52B	Other registered educational non-directory establishment
9	2.53B	Non-government medical and health non-directory establishment
10	2.54B	Community and non-commercial cultural non-directory establishment
11	2.55B	Non-directory establishment in services sector other than education, medical, health, community and non-commercial cultural services

Source Department of Statistics (1987)

than those eligible for schedule 2.51B were surveyed. This sample size of 3 such OAEs was increased to 5 OAEs if the EB had no eligible NDE. If total number of NDEs exceeded 12, a sample of 12 NDEs was selected as follows. First, one NDE was allotted to each schedule type. Next, one more NDE was allotted to different schedules as per the priority order 2.31B, 2.32B, 2.33B, 2.34B, 2.52B, 2.53B, 2.55B, 2.41B and 2.42B. This procedure was repeated until the sample size of 12 NDEs was achieved. The required number of sample NDEs was selected by circular systematic sampling independently for each schedule type. In case of OAEs (other than those eligible for schedule 2.51B), a sample of 3 or 5 OAEs, as appropriate, was selected circular systematically considering all the schedule types together.

- **NSS 41st Round (July 1985–June 1986)**

[List of households and trading enterprises, trade-non-directory establishments and own account enterprises.]

NSS 41st round was another important round that was devoted to the survey on trading enterprises in the unorganized sector. In this round a significant improvement was made in the method of selection of segments i.e., hamlet-groups (HGs)/sub-blocks (SBs) in the large FSUs. This apart, to net adequate number of wholesale trading enterprises whose average turnover is much larger than that of retail trading enterprises, a deeper stratification was made in the urban areas. The sample design of NSS 41st round was, as usual, a stratified two-stage design. The design was integrated to cover directory trading establishments (DTEs) by the CSO and non-directory trading establishments (NDTEs) and own account trading enterprises (OATEs) by the NSSO in the same set of sample FSUs.

In the rural areas census (1981) villages with information on number of enterprises/workers by industry as per the second (1980) EC served as the sampling frame of FSUs (instead of EBs forming the FSUs in 34th round). For the State of Assam, 1977 EC frame was used. Each NSS region within an FOD region generally formed a stratum. In other words, strata were a group districts belonging to the same FOD region x NSS region. Altogether 84 rural strata were formed out of 76 NSS regions at that point of time. In addition, villages missing in the EC frame in certain States formed one separate stratum within each such State. To net adequate number of DTEs and NDTEs in the sample, villages in each stratum were divided into 3 sub-strata[18]: those having at least one DTE (sub-stratum 1), remaining villages with at least one NDTE (sub-stratum 2), and the residual villages of the frame (sub-stratum 3). Rural sample size for a State/UT was first allocated to the strata in proportion to (3DTE + NDTE). Thereafter, out of the stratum level allocation (which was in multiple of 8), an allocation of 8 FSUs was made to sub-stratum 3 and the residual sample size was allocated between sub-strata 1 and 2 in proportion to (3DTE + NDTE). Each stratum x sub-stratum level allocation was in multiples of 8 to facilitate selection of sample FSUs in the form of 2 interpenetrating sub-samples × 4 sub-rounds. The required number of sample FSUs from each sub-stratum x sub-sample was selected by circular systematic sampling with PPS, the size generally being the number of DTEs, number of NDTEs and number of OATEs for sub-strata 1, 2 and 3 respectively after assigning a size of '1' to the villages of sub-stratum 3 having no OATE.[19]

In the urban areas, list of EBs as per the 1980 EC served as the sampling frame of FSUs for 55 class-I cities. In respect of the remaining 170 class-I cities of the country where EBs as per the 1980 EC were not distinctly identifiable and also for all class II to class VI towns, the latest list of UFS blocks was used as the sampling frame for selection of UFS blocks as the FSUs. All the cities and towns in a State/UT were divided into 2 groups, namely, Group A consisting of those having concentration of wholesale traders with the list of such cities/towns provided by the FICCI and also with the inputs of local knowledge of the field staff; and Group B comprising the remaining cities/towns. Altogether 146 class-I cities and 307 classes II to VI cities and towns formed Group A.

Strata were formed within each State/UT adopting, in general, the following broad principle of stratification: (i) each city with the 1981 census population of 500,000 or more formed a separate stratum; (ii) other class-I cities with the use of EC frame; (iii) remaining class-I cities with the use of UFS frame; and (i) all classes II to VI towns together forming another stratum. Total number of urban strata was 101 with the following break-up: 33 strata of cities with population of 500, 00 or more (5 strata with EC frame and 28 strata with UFS frame); 35 strata of other class-I cities (5 with EC frame and 28 with UFS frame); and 33 strata of classes II to VI towns.

[18] There were some minor exceptions in the sub-stratification in certain UTs and in the State of Arunachal Pradesh. For details, see Department of Statistics (1989).

[19] There were some exceptions to this general rule for certain UTs and in case of Arunachal Pradesh for which interested readers may refer to Department of Statistics (1989).

The EBs/UFS blocks with high concentration of wholesale dealers were identified in Group A cities/towns for the purpose of sub-stratification and the following broad principle of sub-stratification was adopted:

Case-I: Cities with the use of EC frame: (i) EBs with high concentration of wholesale dealers (sub-stratum 1); (ii) remaining EBs having at least one DTE (sub-stratum 2); (iii) remaining EBs with at least one NDTE (sub-stratum 3); and (iv) all the residual EBs (sub-stratum 4).

Case-II: Cities and towns where UFS frame was used: (i) UFS blocks with high concentration of wholesale dealers; (ii) remaining UFS blocks identified in the UFS frame as 'bazar area'; and (iii) the remaining UFS blocks of the stratum.

The urban sample size of a State/UT was allotted to the constituent strata in proportion to (3DTE + NDTE) with minor adjustments. Stratum level allocation for class-I cities with the use of EC frame was allocated to 4 sub-strata as follows: a thin sample of 8 FSUs allocated to sub-stratum 4 and the residual sample size of the stratum allocated to sub-strata 1 to 3 in proportion to (3DTE + NDTE) with 4 times weight for sub-stratum 1 to net more wholesale traders in the sample. For the strata with the use of UFS frame, a thin sample of 8 FSUs was allocated to sub-stratum 3 and the remaining sample size of the stratum was allocated to sub-strata 1 and 2 in proportion to the number of UFS blocks with 4 times weight to sub-stratum 1. All the sub-stratum level allocations were adjusted to multiples of 8 to select the sample FSUs in the form of 2 interpenetrating independent sub-samples × 4 sub-rounds with equal allocation to each combination.

For the strata where UFS frame was used, sample UFS blocks were selected circular systematically in the form of 2 independent sub-samples. For the other strata with the use of EC frame, sample EBs were selected by circular systematic sampling with PPS and in the form of 2 independent sub-samples with the size being (3DTE + NDTE) for sub-stratum 1, number of DTEs for sub-stratum 2, number of NDTEs for sub-stratum 3, and number of OATEs for sub-stratum 4 after assigning a size of '1' to the EBs having no OATE as per the frame.

Large sample villages with the approximate present population of 1,200 (600 in case of Himachal Pradesh and Sikkim) or more and large blocks with the same population criterion and/or number of trading enterprises as 100 or more were divided into a suitable number of HGs/SBs for listing of trading enterprises. Before doing so, the compact area within the FSU with the highest concentration of trading enterprises was carved out provided it had at least 10 trading enterprises. This area of concentration was always selected for listing. The remaining area of the FSU was divided into a few HGs/SBs out of which one was selected at random for the listing exercise. Within each FSU (and within each area of concentration/selected HG or SB for larger FSUs), trading enterprises were stratified into 4 s-stage strata (SSS) as follows: OATE—Wholesale (SSS 1); OATE—Others (SSS 2); NDTE—Wholesale (SSS 3); and NDTE—Others (SSS 4). Allocated number of sample enterprises was 14 for FSUs where no area of concentration was carved out (2, 2, 5 and 2 to SSS 1, 2, 3 and 4 respectively). This allocation was larger i.e., 20 in respect of the FSUs where areas of concentration were carved out with an allocation of 14 enterprises to the area of concentration (similar allocations to 4 s-stage strata as was the case for FSUs having no area of concentration) and 6 enterprises to the selected HG/SB with an allocation of 1, 2, 1 and 2 enterprises to SSS 1, 2, 3 and 4 respectively. The required number of enterprises from each SSS was selected circular systematically after arranging the enterprises as per 3-digit National Industrial Classification (NIC) codes.

- **NSS 44th Round (July 1988–June 1989)**

[Survey on living condition of tribals, housing condition and current construction activities, household consumer expenditure.]

Keeping in view the availability of limited data regarding the life of tribal population in the country, a comprehensive survey on the living condition of tribals was carried out in NSS 44th round. In addition, this round covered the subjects of current building construction activity, housing condition and household consumer expenditure ('thin sample'). The survey adopted a stratified two-stage design with 1981 census villages/latest UFS blocks as the FSUs and households (tribal/non-tribal) as the SSUs. An intermediate stage of sampling was selection of a hamlet-group/sub-block randomly (after dividing the FSU into certain number of such segments) for listing purposes in the large FSUs. The sample FSUs were common for different subjects of enquiry. The survey covered the whole country except Ladakh and Kargil districts of Jammu & Kashmir. The rural areas of Nagaland were covered for the first time in this round.

Two household schedules numbered 29.1 and 29.2 were designed to collect information on level of living of tribals and economic activities respectively. For assessing the level of living, the items of information in Schedule 29.1 included details of household consumption, housing condition, social consumption, education, morbidity, inventory of assets and liabilities etc. Schedule 29.2 on economic activities included information on enterprise accounts, land holdings, agricultural operations, dependence on forest, and employment–unemployment. Further, to assess the disparity in living conditions between the tribals and non-tribals, Schedule 29.3 was canvassed to collect the particulars of migration and ownership of land by non-tribals in 'tribal areas'.

As regards the demarcation of rural strata, for the States/UTs with no scheduled tribal population[20] (listed under Article 342 of the Indian Constitution) and also for the States/UTs where the tribal population had a high percentage and they were more or less evenly spread across the districts,[21] each district with 1981 census population less than 1.8 million formed a separate stratum. The districts with larger population were grouped into 2 or more strata by grouping contiguous tehsils similar as far as possible in respect of rural population density and crop pattern. In the remaining States, districts accounting for bulk of the State's tribal population were identified to form strata with concentration of tribal population. Within each such district so identified, tehsils with high concentration of tribal population and together

[20] Chandigarh, Delhi, Daman & Diu, Goa, Haryana, Jammu & Kashmir, Puducherry and Punjab.

[21] Namely, Arunachal Pradesh, Dadra & Nagar Haveli, Lakshadweep Meghalaya, Mizoram, Nagaland and Sikkim.

accounting for at least 70% of the district's total tribal population together formed one stratum, identified as 'Stratum Type 1' or tribal areas. Remaining tehsils of each such district together formed 'Stratum Type 2'. Some districts wholly formed Stratum Type 1. Each of the remaining districts formed a separate stratum by itself. All the strata of States with tribal concentration were identified as type 1 strata. Similarly, all the strata of the States/UTs with no tribal population were identified as type 2 strata.

There were two types of sample villages in this round described as 'general sample villages' and 'special sample villages'. The general sample was drawn from the entire universe where schedules for all the subjects of enquiry namely tribal enquiry, consumer expenditure, building construction and housing condition were carried out. The special sample villages drawn from tribal areas were to augment the general samples for the tribal enquiry. The total number of general sample villages and special sample villages under the central sample was 7,776 and 1,128 respectively. Total urban sample size of FSUs (central sample) was 4,836 urban blocks.

In the urban areas, strata were formed within each NSS region by first grouping towns into 4 population size classes as per the 1981 census population, namely: (i) less than 50,000; (ii) 20,000 to less than 200,000; (iii) 200,000 to less than 1 million; and (i) 1 million and above. Thereafter, towns under size classes (i) and (ii) were further stratified into 2 strata: those having significant tribal population, and the rest. Similarly, all the UFS blocks under town size classes (iii) and (iv) were stratified into 2 strata based on the level of construction activity (i.e., relatively high & low) in the UFS block. Thus, altogether, a maximum of 8 strata were formed within each NSS region.

The State/UT wise allocation of general sample villages, special sample villages and sample blocks was decided by jointly considering the investigator strength, rural–urban distribution of population and the size of tribal population. State/UT level general sample villages were allocated to different strata in proportion to their rural population while special sample villages were allocated to different strata in proportion to tribal population. The urban sample size of a State/UT was allocated to the urban strata in proportion to population with double weights given to the four out of eight strata having significant ST population/high level of concentration activity. NSS region wise sample sizes for rural and urban were in multiples of 8 to ensure equal allocations to 2 sub-samples of FSUs $\times$ 4 sub-round combinations. Sample villages for both general sample and special sample were selected in the form of 2 independent sub-samples by circular systematic sampling with probability proportional to population as per 1981 census (1971 census in case of Assam and a few other districts of some other States where the available list of villages was not satisfactory). Sample blocks were selected circular systematically with equal probability in the form of 2 independent sub-samples.

The hamlet-group/sub-block formation by equalizing population was resorted to as usual in the large villages/blocks and one of them was selected at random for listing purposes. However, in case of large special sample villages, a departure was made in case the tribal population in the village was not evenly distributed over the village. In such cases, the required number of hamlet-groups was formed in a manner that the population of a hamlet-group so formed was nearly half the population of a hamlet-group otherwise formed. From among all such hamlet-groups formed, the one having was largest tribal population was invariably selected (identified as 'area type 1') and more ('area type 2') was selected from the remaining ones at random.

In each sample village, the listed households were arranged in the following order: ST households with means of livelihood as (1) permanent cultivation, (2) shifting cultivation, (3) wage paid manual labour, and (4) others; followed by non-ST households with exactly similar means of livelihood in the same sequence. From the arranged frame of ST households, a combined sample of required number households for Schedules 29.1 and 29.2 was selected circular systematically with equal probability and Schedules29.1 was canvassed in the sample households with odd orders of selection while Schedule 29.2 was canvassed in the other sample households with even orders of selection. An independent sample of non-ST households was selected in the same manner from the arranged frame of non-ST households for canvassing Schedule 29.3. In the urban areas, for canvassing Schedules 29.1 and 29.2, a similar procedure as adopted in rural areas was followed. However, three levels of household MPCE were considered in place of means of livelihood.[22]

- **NSS 48th Round (January 1992–December 1992)**

[Land holdings, livestock holding and debt and investment, household consumer expenditure.]

The main subjects of enquiry of NSS 48th round were land & livestock holdings (LLH) and debt & investment. In fact, this survey on LLH was fifth in the series of similar surveys conducted by the NSSO. In addition, information on household consumer expenditure (HCE) was also collected through a thin sample of households per FSU as part of the annual enquiry on the subject that was introduced in the 42nd round of the NSS. Information on LLH, debt & investment and HCE was collected from the set of sample FSUs. However, data relating to the related subjects were collected from the sample households selected independently from among the listed households.

To reduce the recall error, information on LLH and debt and investment relating to the agricultural year July 1991–June 1992 were collected by visiting each sample household twice during the survey period with a gap of about 4 to 8 months. While the first visit was made during January–August, the second visit took place during the period of September to December 1992. The information on holdings operated during the Kharif season of the agricultural year was collected in first visit while the same on holdings operated during the Rabi season was collected in the second visit.

The survey was planned to cove the whole of the country excluding (i) Ladakh and Kargil districts of Jammu & Kashmir, (ii) interior villages of Nagaland situated beyond 5 km of a bus route, and (iii) villages in Andaman & Nicobar Islands which were inaccessible throughout the year. However, the survey could not be conducted

[22] For further details of sample design including sampling of households and sample size of households per FSU for different schedules of enquiry, interested readers may refer to Department of Statistics (1991) and MoSPI (2004).

in six districts of Jammu & Kashmir namely, Anantnag, Pulwana, Srinagar, Badgam, Baramula and Kupwara, and in the district of Amritsar in Punjab due to unfavourable field conditions.

Mostly, 1981 census list of villages constituted the sampling frame of FSUs i.e., villages in the rural areas. However, in respect of the State of Assam (where 1981 Census was not conducted) and a few districts of other States where the 1981 Census list was incomplete, the 1971 Census list of villages comprised the sampling frame of FSUs. In urban areas, as usual, the latest lists of UFS blocks served as the frame of FSUs.

In rural areas, each district with a rural population of less than 1.8 million according to the 1981 census formed a separate stratum. Districts with larger population were divided into two or more strata by grouping contiguous tehsils, similar as far as possible in respect of rural population density and crop pattern. For the State of Assam, strata were formed exactly in the same manner considering 1971 Census population but with a cut-off of 1.5 million population.

In case of urban, strata were formed within each NSS region based on 1981 census population of towns (1991 census in some of the new towns). Each city with a population of 1 million or more formed a separate stratum. The remaining towns of each NSS region were grouped to form three different strata on the basis of corresponding census population as follows: all towns with population less than 50,000 (stratum 1); all towns with population 50,000 to 199,999 (stratum 2); and all towns with population 200,000 to 999,999 (stratum 3).

The all-India sample size of NSS 48th round was 6,812 FSUs (4,328 villages and 2,484 urban blocks). The State-wise total sample size (determined mainly on the basis of investigator strength and the expected workload per investigator) was allocated to its rural and urban areas in proportion to population with almost double weight to urban. The said rural/urban allocation of a State/UT was further allocated to different strata in proportion to population. All stratum level allocations were adjusted to multiples of 4 to ensure equal number of sample FSUs in each sub-sample and sub-round.

The sample design was broadly a stratified two-stage one with census villages and urban blocks as the FSUs for the rural and urban areas respectively, and households as the SSUs. The selection of villages was done with probability proportional to population (with replacement) while the urban blocks were selected with equal probability without replacement. Sample FSUs for both rural and urban areas were selected in the form of two independent sub-samples.

Large villages and blocks were divided into a suitable number of hamlet-groups and sub-blocks with more or less equal population content. For the listing purposes and sampling of households, two hamlet-groups were selected circular systematically from the large villages whereas only one sub-block was selected with equal probability from the large blocks.

As regards the sampling of households, 2 households were selected by circular systematic sampling (equal probability) from each FSU for collection of data on HCE. For the information on LLH, a random sample of 8 households from each village/ selected hamlet-groups and 9 households from each block/sub-block was selected after due stratification. In the rural areas, listed households with no land possessed or the area of possessed land with less than 0.005 acre were grouped to form stratum 1. All other households were arranged by land possessed in the ascending order to form 3 strata of equal size (in terms of total land possessed) i.e., strata 2, 3 and 4 with a sample allocation of 1, 3, 2 and 2 households respectively to strata 1, 2, 3 and 4 and the required number of sample households was selected circular systematically (with equal probability). In urban, listed households were first grouped into 3 MPCE classes with households having MPCE less than A, between A and B, and B or more. State-level values of A and B were derived based on NSS 43rd round survey on HCE such that roughly 30%, 60% and 10% households belonged to household class 1, 2 and 3 respectively. These MPCE classes were further sub-divided by indebtedness status to form 7 strata in all. A sample of size 1 or 2 households was selected from each sub-stratum. For the sub-strata with allocation of 2 sample households, the samples were drawn circular systematically.

As regards the sample households for the enquiry on debt and investment, in the rural sector, 9 households were selected from each sample village/selected hamlet-groups. For selecting a sample of 9 households, 7 sub-strata were formed. Each of the LLH strata 1 and 2 was further divided into 'indebted' and 'not indebted' groups to form sub-strata 1 to 4. Sub-strata 5 to 7 were formed by first merging LLH strata 3 and 4 and then sub-dividing it into 3 classes i.e., (a) indebted to institutional agencies, (b) indebted to non-institutional agencies alone and (c) not indebted. Finally, an independent sample of 1, 1, 1, 2, 1, 1 and 2 households were selected circular systematically from the sub-strata 1, 2, 3, 4, 5, 6 and 7 respectively. In case of urban FSUs also, a sample of 9 households was selected from each sample urban block/sub-block with same sub-stratification of 7 sub-strata in all as adopted for LLH. Independent samples were selected circular systematically from each of the sub-stratum. The allocated number of sample households to sub-strata 1, 2, 3, 4, 5, 6 and 7 respectively was 1, 1, 1, 1, 2, 2 and 1.

- **NSS 52nd Round (July 1995–June 1996)**

[Survey on health care, participation in education, consumer expenditure and employment and unemployment.]

The subjects of health care, participation in education and household consumer expenditure (short schedule) were taken up in the 52nd round of NSS. A stratified two-stage design with 1991 census villages (1981 census in case of Jammu & Kashmir)/latest UFS blocks as the FSUs and households as the SSUs was adopted. An intermediate stage of sampling was selection of hamlet-groups/sub-blocks in large villages for listing/preparing the frame of households. For Kerala, list of panchayat wards was used as the frame for sampling of panchayat wards as the FSUs. For the State of Nagaland, villages situated within 5 km. of a bus route served as the frame. Similarly, for Andaman & Nicobar Islands, the list of accessible villages was used as the sampling frame. Three different schedules of enquiry on the three subjects were canvassed in different sets of sample households. However, the sample FSUs were common for all the schedules of enquiry.

Each district within an NSS region was considered as a separate stratum in rural areas. However, if the 1991 census population of the district was 2 million (1.8 million population as per 1981 census for Jammu & Kashmir) or more, the district was divided into two or more strata, by grouping contiguous tehsils to form strata. In Gujarat, in the case of districts extending over more than one NSS region, the part of a district falling within each NSS region formed a separate stratum. Urban strata were formed by grouping cities and towns as per their 1991 census population as follows: all towns with population less than 50,000 (stratum 1); towns with population between 50,000 and 200,000 (stratum 2); all towns with population between 200,000 and 1,000,000 (stratum 3); and each million plus city forming a separate stratum by itself (stratum 4, 5 and so on).

All-India sample size (central sample) was fixed at 13,000 FSUs for rural and urban areas combined. This was allocated to States/UTs considering the distribution of available investigator strength and population of respective States/UTs. State/UT level total sample size was allocated between rural and urban in proportion to population with double weight to urban. Total all-India sample size was 7,888 villages and 5,112 urban blocks. Rural or urban total sample size of FSUs of a State/UT was allocated to different strata in proportion to their population as per the census with stratum-wise allocations adjusted to multiples of 8 to the extent possible (else multiples of 4) to allocate equal number of FSUs to each sub-sample x sub-round combination (with 2 sub-samples and 4 sub-rounds of 3 months each). Sample FSUs from each stratum for both rural and urban areas was selected by circular systematic sampling with equal probability and in the form of 2 independent sub-samples. However, a departure was made for sampling of villages in Arunachal Pradesh where a cluster sampling was followed. The nucleus villages were selected circular systematically with equal probability and a cluster of 4 to 6 villages was formed around each nucleus village for the survey.

Large rural FSUs were divided into 4 or more hamlet-groups (HGs) and large FSUs in urban areas were divided into 2 or more sub-blocks (SBs). Thereafter, 2 HGs in rural and 1 SB in urban was selected randomly from among the HGs/SBs formed for listing and sampling of households. Finally, from each selected FSU or 2 HGs combined/1 SB for large FSUs, a sample of 10 households, 6 households and 4 households respectively were selected for collecting data on health care, education and household consumer expenditure (HCE) with due stratification for the first two subjects of enquiry as follows (Box 3.3):

The required number of sample households for each subject of enquiry (and from each stratum in case of health care and education) was selected circular systematically with a random start after arranging the listed households in the respective frames by their means of livelihood. The means of livelihood of the households were described as follows: (a) rural samples: self-employed in non-agriculture (code 1), rural labour (code 2), and others (code 9); and (b) urban samples: self-employed (code 1), regular wage/salary earning (code 2), casual labour (code 3), and others (code 9). Such an arrangement was to ensure proportionate representation of households with different means of livelihood in the ultimate sample.

Box 3.3 Method of stratification of households, NSS fifty-second round

Health care		Education		HCE
Stratum and its composition	Number of sample households	Stratum and its composition	Number of sample households	Number of sample households
1: Households with at least one child of age '0' year	2	1: Households with at least one member of age group 5–24 years and studying in post-primary level	3	4
2: Of the remaining households, those reporting any case of hospitalisation during the last 365 days	2	2: Remaining households	3	
3: Remaining households	6	–	–	

Source Department of Statistics (1998)

- **NSS 55th Round (July 1999–June 2000)**

[Consumer expenditure, employment and unemployment, informal sector enterprises.]

(Highlighting changes in relation to NSS 50th round: July 1993–June 1994)

NSS 55th round was the sixth quinquennial round of the survey on employment–unemployment (E-U) and household consumer expenditure (HCE).[23] The immediate previous i.e., fifth quinquennial round was the 50th round of the NSS. Some important features of NSS 55th round were as follows:

(i) Unlike the other quinquennial rounds, this round also covered informal non-agricultural enterprises (other than those engaged in industrial categories of mining & quarrying, electricity, gas and water supply).

(ii) Keeping in view the respective sizes of the E-U and HCE schedules of enquiry, this round canvassed the two schedules of enquiry on E-U and HCE in two different/independent sets of households unlike in the previous surveys where details of employment–unemployment particulars and household consumer expenditure were collected from the same sample households. This change was primarily done to reduce the time of data collection in the selected households and this procedure continued in the succeeding quinquennial rounds also.

(iii) Information on consumption of food, pan, tobacco and intoxicants was collected independently for two different reference periods of last 7 days and

[23] In all, there have been nine NSS rounds of the quinquennial surveys on E-U and HCE with the corresponding NSS round numbers being 27, 32, 38, 43, 50, 55, 61, 66 and 68. Refer to Statement 2.1 of chapter 2 for the corresponding survey periods.

last 30 days from the same households.[24] In the past, this information was collected for only the reference period of last 30 days.

(iv) In contrast, information for items of clothing, footwear, education, medical (institutional) and durable goods was collected for only one reference period of last 365 days. This information was collected for two different reference periods of last 30 days and last 365 days in the earlier quinquennial rounds of HCE.[25]

(v) Another salient feature of this round was the adoption of rotational sampling scheme for the first time in the NSS for the purpose of collection of employment–unemployment data from central sample[26] only. Under this scheme, one sub-sample of the sample FSUs of each sub-round was revisited in the subsequent sub-round. From each such FSU, sample households visited in the previous sub-round were revisited in the subsequent sub-round for collecting employment and unemployment particulars.

A stratified design, with villages (panchayat wards in case of Kerala) and UFS blocks as the FSUs and households (for the surveys on E-U & HCE)/enterprises (for the survey on informal non-agricultural enterprises) as the ultimate stage units, was

---

[24] Although the instruction was to collect the data for the two reference periods independently from a sample household, the suggested method created controversy with some scholars commenting that the responses on different items of consumption based on the one reference period (particularly, the 7-day) had affected /influenced the responses as per the other reference period, affecting the comparability of the estimates as per this survey with those of the past. In view of this criticism, this method of data collection was dispensed with in the next quinquennial surveys.

[25] As regards the reference periods for the collection of data on HCE, there have been certain changes over NSS rounds with some explained under (iii) and (iv) above. Readers may note the following further clarifications in this regard (MoSPI, 2001b): 'The annual series of consumer expenditure surveys, up to the 49th round, used a uniform reference period of "last 30 days" for all items of consumption. In the bigger surveys of the quinquennial series, an additional reference period of "last 365 days" was used for some items of consumption – particularly, clothing, footwear and durable goods—but most results were tabulated using the 'last 30 days' data. During the 51st to 54th rounds, one-half of the sample of households was surveyed through schedule type I, which had a reference period of 30 days for all items. In the other half of the sample, a schedule (schedule type 2) with different reference periods for different items was tried out on an experimental basis. The reference periods for schedule type 2 were: Food, pan tobacco and intoxicants: last 7 days; Fuel & light, miscellaneous goods &services, and medical (non-institutional): last 30 days; Educational, medical (institutional), clothing, footwear and durable goods: last 365 days. During 55th round of NSS, information on consumption of food, pan, tobacco & intoxicants was collected independently for two different reference periods of 7 days and 30 days from the same households. Another important feature of this round was that information in respect of consumption of clothing, footwear, durable goods, education and medical (institutional) services was collected only for one reference period of "last 365 days". In some of the quinquennial rounds, information on these items was collected for two different reference periods - last 30 days and last 365 days - from the same households'. (MoSPI, 2001b). For the reference periods adopted in the NSS 68th round (July 2011–June 2012), the readers may refer to the discussion that follows in this chapter under the caption 'NSS Sixty-eighth Round (July 2011–June, 2012)'.

[26] 'Central sample' comprises the sample FSUs covered by the NSSO employing its manpower. The counterpart is the 'state sample' consisting of the matching sample FSUs covered by the States/ UTs participating in the NSS.

adopted. Large FSUs were as usual segmented and the listing exercise was carried out in the selected segments only.

List of 1991 census villages (panchayat wards in case of Kerala)/1981 list of census villages for the State of Jammu & Kashmir and the latest of UFS blocks served as the sampling frame of FSUs. All the uninhabited villages of the country as per 1991 Census, interior villages of Nagaland which were situated beyond 5 km of a bus route and inaccessible villages of Andaman & Nicobar Islands were left out of the survey coverage.

As regards the rural stratification, each district was generally considered as a separate stratum by itself. Each district with rural population of 2 million or more as per 1991 Census (1.8 million or more as per 1981 Census in case of Jammu & Kashmir) was, as usual, split into a number of strata. However, 2 special strata were formed at the State/UT level by taking out smaller/larger villages from the general strata as follows: all FSUs with population between 1 to 100 (stratum 1); and FSUs with population more than 15,000 (stratum 2). NSS 50th round also adopted a similar method of stratification (with 1981 list of villages generally used as the frame of FSUs for rural areas) with no such special strata formed in that round.

Urban strata were formed within each NSS region in both NSS 55th and 50th rounds. To capture adequate number of enterprises in NSS 55th round, UFS blocks with type of area as hospital area (HA) or industrial area (IA) or bazaar area (BA) in different size classes of towns were grouped together to form one type of stratum. As against the same, in NSS 50th round, UFS blocks belonging to 'affluent areas' of the larger cities with population of 400,000 or above were grouped to form one type of stratum with a view to capturing adequate number of households from the affluent section of the society. The information on such affluent areas was supplied by the regional/sub-regional offices of FOD, NSSO. Boxes 3.4.1 and 3.4.2 give the details of strata formed within each NSS region in the respective rounds.

The all-India sample size (central sample) in NSS 55th round was 10,384 FSUs (Rural: 6,208; Urban: 4,176). This was much lower than the sample size (12,040 FSUs) adopted in 50th round (Rural: 7,248; Urban: 4,792). State/UT x rural/urban x stratum wise allocations were finalized keeping in view the proportions of population

Box 3.4.1 Stratification of urban areas in NSS 55th round

Stratum number	Composition of the stratum
1, 3, 5	HA/IA/BA blocks of each single city with population 1 million or more
2, 4, 6	Other blocks of each single city with population 1 million or more
7	HA/IA/BA blocks of all towns with population 50,000 or more but less than 1 million
8	Other blocks of all towns with population 50,000 or more but less than 1 million
9	HA/IA/BA blocks of all towns with population less than 50,000
10	Other blocks of all towns with population less than 50,000

Source MosPI (2001b)

Box 3.4.2 Stratification of urban areas in NSS 50th round

Stratum number	Composition of the stratum
1	All towns with population less than 50,000
2	All towns with population between 50,000 to less than 200,000
3	All towns with population between 200,000 to less than 400,000
4	Towns with population between 400,000 to less than 1 million: Blocks in affluent areas
5	Towns with population between 400,000 to less than 1 million: All other blocks of the towns
6	A single city (if any) with population 1 million or more: Blocks in affluent areas
7	A single city (if any) with population 1 million or more: All other blocks of the city
8	Another single city (if any) with population 1 million or more: Blocks in affluent areas
9	Another single city (if any) with population 1 million or more: All other blocks of the city

Source Department of Statistics (1996)

with double weights given to strata consisting of HA/IA/BA blocks in 55th round and double weights given to strata containing blocks of affluent areas in 50th round.

Sample FSUs in NSS 55th round were divided equally into 4 sub-rounds. Stratum level allocations of a sub-round were in even numbers for selection of sample FSUs in the form of 2 independent sub-samples. Sub-sample numbers were 1 & 2 for sub-round 1; 3 & 4 for sub-round 2; 5 & 6 for sub-round 3 and 7 & 8 for sub-round 4. For each sub-round, sample FSUs from each stratum were selected in the form of 2 independent sub-samples by circular systematic sampling with probability proportional to population for all rural strata other than stratum 1 and with equal probability for rural stratum 1 as well as all urban strata.

The large FSUs were divided into a minimum of 5 hamlet-groups (HGs for rural)/sub-blocks (SBs for urban) of roughly equal population content by taking into consideration the approximate present population and approximate total number of non-agricultural enterprises in the FSU.[27] From among those, the HG/SB having maximum concentration of non-agricultural enterprises was selected with certainty (termed as 'segment 1') for listing purposes. From the remaining HGs/SBs, 2 more were selected circular systematically and they were merged/identified as 'segment 2' for a combined listing of households and enterprises. As against this procedure of HG/SB selection, in NSS 50th round, 2 HGs were selected circular systematically from large villages and 1 SB was selected randomly from the large blocks.

For the purpose of sampling of households for both E-U and HCE, all the households listed in the FSU/segment (both rural & urban) were stratified into 2 strata:

[27] See MoSPI (2001b) for further details.

'affluent households' (second-stage stratum 1) and the rest of the households (second-stage stratum 2). In the rural areas, a household was classified as 'affluent' if it owned certain items like motor car/jeep, colour TV, telephone, etc. or owned land/livestock in excess of certain limits. However, in case of urban FSUs, the households with MPCE exceeding certain limit were treated as 'affluent' households. The allocated number of sample households per FSU for canvassing each of Schedule 10 (Schedule 10.1 for revisit) on E-U and Schedule 1.0 on HCE was as follows:

Segment	Second-stage stratum		
	1	2	Combined
FSUs with no HG/SB formation (Identified as Segment 1)			
1	2	10	12
FSUs with HG/SB formation			
1	1	3	4
2	1	7	8

Source MosPI (2001b)

The FSUs of sub-sample 1, sub-sample 3 and sub-sample 5 were revisited during sub-round 2, sub-round 3 and sub-round 4 respectively. In the FSUs of these revisit samples, all the households where schedule 10 was canvassed during the previous sub-round were revisited for canvassing schedule 10.1. For each of E-U and HCE, the required number of sample households from each second-stage stratum were independently selected by circular systematic sampling with equal probability. For the purpose of systematic sampling, households in the frame of second-stage stratum 2 were arranged by means of livelihood x land possessed classes for rural samples and by means of livelihood x MPCE classes for urban samples. In NSS 50th round, a more or less similar approach was adopted for stratification and sampling of households.[28]

For sampling of informal non-agricultural enterprises listed enterprises in a segment (both rural & urban) are stratified into 12 strata based on their broad industry group (6 in number) x enterprise class (2 in number). Broad industry groups were: manufacturing—1, construction—2, trade & repair services—3, hotels & restaurants—4, transport, storage & communication—5 and other service sector—6. Enterprise class of an enterprise was '1' for Own Account Enterprises (i.e., those without any hired worker) and '2' for Others having hired workers. Thus, there are 12 possible strata of various combinations of broad industry groups and enterprise classes. A total sample size of 24 enterprises per FSU was allocated for survey i.e., 2 from each of the 12 strata from FSUs without HG/SB formation and 1 from each of the 12 strata for each of segment 1 and 2. The required number of sample enterprises from each stratum (i.e., segment x broad industry group x enterprise class) was selected circular systematically after arranging the listed enterprises in the ascending order of industry (NIC) 2-digit codes (3-digit codes for hotels & restaurants) before sampling if 2 or more enterprises were to be selected from a given stratum.

[28] For further details, refer to Department of Statistics (1996).

3.4 The Third Phase (2000–01 to 2016–17): NSS Round Numbers 56–74

- **NSS 58th Round (July 2002–December 2002)**

[Village facilities, slum, disabled persons, housing condition, consumer expenditure.]

NSS 58th round was another important round of six months' duration covering the subjects of slum, disability, housing condition, consumer expenditure (thin sample) and village facilities. As regards the survey on slums, the survey in this round was in fact the third in the series of nationwide surveys on this subject conducted as part of the NSS with the previous surveys carried out in NSS 31st round (July 1976–June 1977) and 49th round (January–June 1993). The 31st round survey was confined to only the class—I towns of 1971 census having population of 100,000 or more and two class—II towns namely Shillong and Pondicherry. NSS 49th survey, however, covered entire rural and urban areas of the country. For the purpose of the survey, a 'compact area with a collection of poorly built tenements, mostly of temporary nature, crowded together usually with inadequate sanitary and drinking water facilities in unhygienic conditions' was considered as 'slum' For the purpose of the survey, such an area was considered as 'undeclared slum', if at least 20 households lived in that area. With the slum basically being an urban concept, it was decided to cover only urban slums in NSS 58th round. For each slum in the sample UFS blocks, particulars about the slums were collected by canvassing Schedule 0.21. Thus, no household schedule was relevant for the survey on slums.

A stratified multi-stage design was adopted for the survey with FSUs as the 1991 census villages (panchayat wards for Kerala; 1981 census villages for Jammu & Kashmir) in the rural areas and the latest UFS blocks in the urban areas. The USUs (ultimate stage units) were households. As usual, an intermediate stage of sampling was the selection of HGs/SBs in the large FSUs. The sample FSUs were common for all the subjects of enquiry. However, sample households for different subjects of enquiry were selected independently with due stratification of the households. The survey covered the entire country excluding (i) Leh and Kargil districts of Jammu & Kashmir, (ii) interior villages of Nagaland situated beyond 5 kms of the bus route and (iii) villages in Andaman and Nicobar Islands which remained inaccessible throughout the year. The survey period of 6 months was divided into 2 sub-rounds of 3 months each and equal number of sample FSUs was allotted for survey in each sub-round.

The all-India total sample size of FSUs (rural and urban combined) for the central sample was fixed at 8,338 (3,552 FSUs for urban) which was further allocated to different States and UTs in proportion to provisional population as per Census 2001, subject to the availability of investigators to tackle the workload. State/UT level total sample size was further allocated between rural and urban areas of the State/UT in proportion to provisional population as per Census 2001 with double weight given to urban.

Each district was normally treated as a separate stratum in the rural areas. However, if the provisional 2001 census population of the district was 2.5 million or more, the district was divided into two or more strata of more or less equal population contents as per population census 1991 by grouping contiguous tehsils. In the State of Gujarat where some districts cut across NSS region boundaries, the part of each such district falling within an NSS region constituted a separate stratum. Before forming such general strata, very small and large FSUs were taken out to form 2 special strata at the State/UT level as follows: those with 1991 census population between '0' to 50 (stratum 1) and FSUs with population more than 15,000 (stratum 2).

In the urban areas, strata were formed within each NSS region based on population size class of the towns as per 1991 census as follows:

Stratum 1: All towns with population less than 0.1 million;
Stratum 2: All towns with population 0.1 million to less than 0.5 million;
Stratum 3: All towns with population 0.5 million to less than 1 million; and.
Stratum 4,5,6, …: Each town with population 1 million or more.

To capture adequate number of households residing in slum areas, each urban stratum was sub-stratified into 2 sub-strata, namely, all UFS blocks of the stratum with area type as 'slum area' (sub-stratum 1), and the remaining UFS blocks of the stratum (sub-stratum 2). No sub-stratification was resorted to in the rural areas.

The rural and urban sample size of a State/UT was allocated to different strata in proportion to population with such allocations adjusted to multiples of 4 to facilitate equal allocations to each of the combination of 2 sub-rounds × 2 sub-samples. The stratum level allocations for urban areas was further allocated to 2 sub-strata in proportion to the number of UFS blocks with double weight to sub-stratum 1 subject to a minimum sample size of 2 or 4 to sub-stratum 1 depending upon whether the stratum level allocation was 4 or greater than 4. Sub-stratum level allocations in the urban sector were made even numbers.

From each stratum/sub-stratum, sample FSUs were selected in the form of two independent sub-samples. For special stratum 2 and all the general strata of rural areas, FSUs were selected by probability proportional to size with replacement (PPSWR) where size was the 1991 census population. For urban areas and special stratum 1 of rural areas, FSUs were selected by simple random sampling without replacement (SRSWOR).

The large sample FSUs with approximate population on the date of survey as 1,200 or more[29] was divided into a minimum of 3 HGs/SBs by more or less equalizing population. For large urban blocks, the SB having slum dwellers, if any, was selected with probability 1 and it was identified as segment 1. In case there was more than one such SB, the one having maximum number of slum dwellers was identified as segment 1. One more SB was selected by simple random sampling (SRS) from the remaining SBs of the block and the same was identified as segment 2. For large blocks

[29] A reduced threshold of 600 or more was considered for this purpose (as has been the practice) in case of rural areas of Himachal Pradesh, Sikkim and Poonch, Rajouri, Udhampur and Doda districts of Jammu and Kashmir and Idukki district of Kerala where habitation pattern causes difficulty in listing.

having no slum areas, two SBs were selected by SRSWOR and were combined to form segment 2. From large villages two HGs were selected by SRSWOR and were combined to form segment 2. The villages without HG formation were also identified as segment number 2. The selected segments were considered separately for listing and selection of sample households.

The listed households were independently stratified for canvassing of schedules pertaining to different subjects of enquiry i.e., housing condition (Schedule 1.2), HCE (Schedule 1.0) and disability (Schedule 26). An independent sample of 12 households for Schedule 1.2, 4 households for Schedule 1.0 and 12 households for Schedule 26 per FSU was selected with the stratification and sample allocation as under (see Box 3.5):

From each second-stage stratum, the required number of sample households was selected by SRSWOR.

- **NSS 59th Round (January–December 2003)**

[Situation assessment survey of farmers, land and livestock holdings, debt and Investment, consumer expenditure.]

In order to study the well-being of farmer households, a Situation Assessment Survey (SAS) of farmers' households was carried out in the NSS 59th round as per the request of the Ministry of Agriculture and Farmers Welfare. Schedule 33 was canvassed in a randomly drawn sample of farmer households. In addition, the other subjects of enquiry and the corresponding schedules in this round included land and livestock holdings (Schedule 18.1), debt and Investment (Schedule 18.2), and consumer expenditure (Schedule 1.0). By farmer household, the survey meant a household that possessed some land and was engaged in some farming activity during the last 365 days. While the SAS covered only the rural areas of the country, all the other subjects covered both rural and urban areas.

The survey covered the whole of the Indian Union except (i) Leh (Ladakh) and Kargil districts of Jammu & Kashmir, (ii) interior villages of Nagaland situated beyond 5 kms of the bus route and (iii) villages in Andaman and Nicobar Islands which remained inaccessible throughout the year. A stratified multi-stage design was adopted for the survey with the 1991 census villages (panchayat wards for Kerala and 1981 census villages for Jammu and Kashmir) and the latest UFS blocks as the FSUs and households as the USUs. In the large FSUs, HGs/SBs constituted the intermediate i.e., second-stage sampling units.

The survey period was divided into two sub-rounds with equal number of sample FSUs allotted to each. With the exception of consumer expenditure, all the other enquiries were carried out by visiting each sample household twice, the first one (Visit 1) during January to August and the other (Visit 2) during September to December. Accordingly, each sample FSU (village or block) was visited twice by the investigator to whom it was allotted. Sub-round 1 consisted of the first half of the survey period of each visit while sub-round 2 consisted of the remaining period.

The stratification of rural FSUs in this round was exactly similar to that adopted in the previous round as discussed. Urban strata were formed within each NSS region

Box 3.5 Stratification and sample allocation of households in NSS fifty-eighth round

Area	Second-stage stratum no.	Stratum composition	Number of sample households per FSU		
			FSUs without segment selection or large FSUs with segment 2 only	Large urban FSUs having both segments 1 and 2	
				Segment 1	Segment 2
Schedule 26 (Survey of Disabled Persons)					
Rural/Urban	1	HHs having at least one person with mental disability	4	2	2
	2	Among the remaining HHs, those having at least one person with speech/ hearing/visual disability	4	2	2
	3	Among the remaining HHs, those having at least one person with locomotor disability	4	2	2
Schedule 1.2 (Housing Condition)					
Rural	1	Households (HHs) having pucca dwelling structure	4	–	–
	2	HHs having semi-pucca dwelling structure	4	–	–
	3	Remaining HHs	4	–	–
Urban	1	HHs having MPCE of top 10% of urban population	4	2	2
	2	HHs having MPCE of middle 60% of urban population	4	2	2
	3	HHs having MPCE of bottom 30% of urban population	4	2	2

(continued)

Box 3.5 (continued)

Area	Second-stage stratum no.	Stratum composition	Number of sample households per FSU		
			FSUs without segment selection or large FSUs with segment 2 only	Large urban FSUs having both segments 1 and 2	
				Segment 1	Segment 2
Schedule 1.0 (Household Consumer Expenditure)					
Rural	1	HHs having pucca dwelling structure	2	–	–
	2	Remaining HHs	2	–	–
Urban	1	HHs having MPCE of top 10% of urban population	2	1	1
	2	Remaining HHs	2	1	1

Source MoSPI (2003)

by grouping towns as per their population size classes based on Population Census 2001:

Stratum 1: All towns with population less than 50,000;
Stratum 2: All towns with population 50,000 or more but less than 200,000;
Stratum 3: All towns with population 200,000 or more but less than 999,999; and.
Stratum 4, 5, 6,…: Each city with population 1 million or more.

All-India sample size of FSUs was fixed at 10,608 (Rural: 6,784 and Urban: 3,824) which was allocated to States/UTs x rural/urban x strata keeping in view their shares in population with 1.5 weights given to urban areas at the State/UT level. The earlier practice of giving double weight to urban areas was modified in this round keeping in view the fact that 'two important subjects of enquiry (Land and Livestock Holdings and Situation Assessment Survey of Farmers) were rural-based and there had been considerable growth in urban population over the years'. From all the rural strata except stratum 1, sample FSUs were selected with PPSWR, size being the population as per Population Census 1991. From stratum 1 of the rural areas and from all the urban strata, selection was done using SRSWOR. Sample FSUs from all the strata were drawn in the form of 2 independent sub-samples in both rural and urban areas.

Large villages/blocks having approximate present population 1,200 or more were divided into a minimum number of 3 HGs/SBs of roughly equal population content and two of them were selected by SRSWOR for independent listing and sampling of households in the respective HGs/SBs.

Sample households were independently selected for canvassing schedules pertaining to different subjects of enquiry after due stratification of the listed households (HHs). Box 3.6.1 gives the details of stratification and sample allocation per FSU for canvassing Schedules 33 and 1.0 in NSS 59th round:

Box 3.6.1 Stratification and sample allocation of households for surveys on situation assessment survey of farmers and consumer expenditure in NSS fifty-ninth round

Second-stage stratum	Stratum composition	Number of sample HHs per FSU^
Schedule 33 (SAS of farmers): Rural Only		
1	HHs possessing land less than 0.005 ha and engaged in farming activity during the last 365 days	2
2	HHs possessing land equal to or more than 0.005 ha but less than X and engaged in farming activity during the last 365 days*	2
3	HHs possessing land equal to or more than X but less than Y and engaged in farming activity during the last 365 days*	2
4	HHs possessing land equal to or more than Y and engaged in farming activity during the last 365 days*	2
Schedule 1.0 (Consumer Expenditure): Rural/Urban		
1	HHs possessing land < Y (rural) or MPCE < C (urban)**	2
2	Other HHs	2

Note *Values of X and Y were estimated at the State/UT level from the data of NSS 48th round survey of land holdings such that among the households having land area of 0.005 hectare or more, 40% of households possessed land area less than X, 40% possessed land area between X and Y and 20% possessed land area greater than Y.
**The area of land Y was determined for each State/UT from the data of NSS 48th round survey such that among the households having land area of 0.005 hectare or more, 80% of the households possessed land less than Y. Similarly, using NSS 55th round data, the MPCE value of C was determined at State/UT level such that 90% of the households in the State/UT had MPCE less than C.
^ The said sample allocation for each second-stage stratum was 1 each for the two selected HGs/SBs in case of large FSUs with HG/SB formation.
Sources MoSPI (2005a, c)

The second-stage stratification for canvassing Schedules 18.1 and 18.2 were related. While 4 s-stage strata (SSS) i.e., land and livestock holdings (LLH) strata were formed to select households for Schedule 18.1 by taking into account the area of land possessed by the households for rural areas or the value of MPCE of the households in case of urban areas, a total number of 7 SSS were formed for the purpose of canvassing Schedule 18.2 by jointly considering the area of land possessed (rural)/ MPCE (urban) and status of indebtedness of the households. Box 3.6.2 gives the details of stratification of listed households and sample allocations to various strata in respect of Schedules 18.1 and 18.2:

For each of the four schedules of enquiry, the required number of sample households from each stratum was selected by SRSWOR.

NSS 64th Round (July 2007–June 2008)

[Migration, Employment–Unemployment, Education, Household Consumer Expenditure.]

Box 3.6.2 Stratification and sample allocation of households for surveys on land & livestock holdings and debt & investment in NSS fifty-ninth round

Second-stage stratum	Stratum composition	Number of sample HHs per FSU^
Schedule 18.1 (Land & Livestock Holdings): Rural*		
1	HHs with land less than 0.005 hectare	2
2	HHs with land equal to or more than 0.005 hectares but less than X	2
3	HHs with land equal to or more than X but less than Y	2
4	HHs with land equal to or more than Y	2
Schedule 18.1 (Urban)*		
1	HHs with MPCE less than A	2
2	HHs with MPCE equal to or more than A but less than B	2
3	HHs with MPCE equal to or more than B but less than C	2
4	HHs with MPCE equal to or more than C	2
Schedule 18.2 (Debt & Investment): Rural/Urban		
1	Indebted HHs belonging to LLH Stratum 1	2
2	Non-indebted HHs belonging to LLH Stratum 1	2
3	Indebted HHs belonging to LLH Stratum 2	2
4	Non-indebted HHs belonging to LLH Stratum 2	2
5	HHs indebted to institutional agencies with or without being indebted to non-institutional agencies belonging to LLH Strata 3 & 4	2
6	HHs indebted to non-institutional agencies only belonging to LLH Strata 3 & 4	2
7	Non-indebted HHs belonging to LLH Strata 3 & 4	2

Notes * Values of X and Y were estimated at the State/UT level based on NSS 48th round survey of land holdings such that among the households having land area of 0.005 hectare or more, 40% of households possessed land area less than X, 40% possessed land area between X and Y and 20% possessed land area greater than Y.
** By utilizing the data on HCE as per NSS 55th round, values of AB B and C were estimated at the State/UT level such that 30% of the households had MPCE less than A, 30% had MPCE more than or equal to A but less than B, 30% had MPCE more than or equal to B but less than C and the remaining 10% of the households had MPCE more than or equal to C.
^ The said sample allocation for each second-stage stratum was 1 each for the two selected HGs/SBs in case of large FSUs with HG/SB formation.
Sources MoSPI (2005b, 2006b)

As a part of the surveys on employment–unemployment (E-U), the data on migration particulars were collected in many NSS rounds starting with the 9th round (May–August 1955) followed by 11th to 13th rounds and subsequently as part of the quinquennial surveys on E-U during NSS 38th round (January–December 1983), 43rd round (July 1987–June 1988) and 55th round (July 1999–June 2000) and also thereafter as part of NSS 64th round (July 2007–June 2008) which covered the subjects of education and HCE apart from E-U and migration. Finally, PLFS: 2020–21, through a separate block in the questionnaire, also collected certain migration particulars of the household members. Apart from these rounds devoted to E-U, information on migration was also collected during NSS 18th round (February 1963–January 1964), NSS 28th round (October 1973–June 1974) and NSS 49th round (January–June 1993) which also covered other subjects of enquiry.

NSS 64th round was a comprehensive survey on migration collecting information about the migration particulars of the household members as well certain information about the out-migrants, 'seasonal short-term migrants' and return migrants. For the purpose of the survey, a household member was, as usual, treated as a migrant if his or her last usual place of residence (UPR),[30] anytime in the past, differed from the present place of enumeration. Short-term migrants were those who had undertaken short-term movements without changing their UPR. Return migration related to a situation where a migrant returned back to their earlier UPR.

The information collected during NSS 64th round relating to migrants, inter-alia, included the location of their last UPR, reason for migration, period elapsed since migration, usual activity pursued at the time of migration and whether he or she was a return migrant. Items of information relating to out-migrants included 'present place of residence, reason for migration, period since leaving the household, whether presently engaged in any economic activity, whether sent remittances, number of times and amount of remittances sent during last 365 days'. For the short-term migrants who had stayed away, during the last 365 days, for employment or in search of employment for a period of 30 days to 6 months, information on 'number of spells of duration of 15 days or more, destination for longest spell and industry of work, if worked' were collected. For further details, readers may refer to NSS Report Number 533, Migration in India, 2007–2008, NSS 64th Round.

NSS 64th round survey had an all-India coverage excluding Ladakh and Kargil districts of Jammu & Kashmir (for central sample), the interior villages of the State of Nagaland located beyond 5 kms of the bus route and the villages of Andaman and Nicobar Islands which remained inaccessible throughout the year. As part of the central sample, the survey was conducted in a random sample of 7,921 villages and 4,668 UFS blocks covering 79,091 rural households and 46,487 urban households of the country.

For the selection of sample FSUs, as usual, the survey adopted a stratified multi-stage design with villages/UFS blocks comprising the FSUs and households as the USUs. The list of census 2001 villages (panchayat wards for Kerala) in case of rural areas and the latest list of UFS blocks for the urban areas served as the sampling frame of FSUs. However, for the newly declared towns and out growths as per census 2001 for which UFS was not done, each individual town/OG was considered as an

[30] Usual place of residence was defined as a place (i.e., village or town) where the person had stayed continuously for a period of six months or more.

FSU and all such towns/OGs also formed a part of the overall sampling frame of FSUs for the urban areas.

For both rural and urban areas, each district was considered as a basic stratum. However, in case of urban, if the district had any town with a population of one million or more as per census 2001, each of such towns/cities formed a separate stratum by itself and all the remaining towns of the district together formed another separate stratum within the district. Each rural/urban stratum was sub-stratified into n/4 sub-strata with 'n' denoting the number of sample FSUs for a stratum, after arranging the villages/towns in ascending order of population as per the frame such that each sub-stratum had roughly equal population in case of rural areas and equal number of FSUs in case of urban. For the urban strata of towns/cities with a population of one million or more, the said arrangement was based on Investigator Unit number to which the UFS blocks belonged × UFS block numbers in their ascending order within an Investigator Unit.

The overall all-India sample size of FSUs was allocated to States/UTs in proportion to census 2001 population with a minimum allocation to each State/UT. The State/UT level allocation was allotted between the rural and urban areas of the State/UT in proportion to population with 1.5 weights to urban areas[31] by adhering to the constraint that the urban allocation for the bigger States like Maharashtra, Tamil Nadu, etc., did not exceed the rural sample size. A minimum sample of 8 FSUs was allocated to each state/UT, separately, for rural and urban areas. The respective State/UT x rural/urban allocation of FSUs was further allocated to different strata in proportion to the respective stratum population as per census 2001. Allocations at stratum level were adjusted to a multiple of 4 with a minimum sample size of 4. Finally, a sample of 4 FSUs was allocated to each sub-stratum. From each sub-stratum, the sample FSUs in rural areas were selected with PPSWR, size being the population as per census 2001. For urban sector, the sample FSUs were selected by SRSWOR. From each sub-stratum, the sample FSUs were drawn in the form of two independent sub-samples for both rural and urban areas.

The large sample FSUs having an approximate present population of 1,200 or more was divided into a minimum number of 3 segments,[32] called hamlet-groups (HG) for rural areas and sub-blocks (SB) for urban areas by more or less equalizing population, and a sample of 2 segments was selected by SRSWOR to prepare the frame of USUs within the selected segments. The frame of households and the sampling of households were independently done for each of the two selected segments.

[31] While allocating the State/UT level overall sample size of FSUs between rural and urban areas by considering respective rural and urban populations, the NSS generally gives a higher weightage of 1.5 or 2 to urban areas (resulting in the oversampling of urban areas) since the urban areas and the USUs therein are likely to have a higher variability.

[32] The actual number of segments depended on the approximate population in the entire FSU with a larger number of segments formed in the bigger FSUs having more population. The above population threshold was taken as 600 for the purpose of segment formation in rural areas of Himachal Pradesh, Sikkim and Poonch, Rajouri, Udhampur, Doda districts of Jammu and Kashmir as well as in Idukki district of Kerala which had hilly terrains.

Finally, for the purpose of sampling of households to canvass the schedule on E-U and migration, all the listed households within the FSU/selected segments were divided into 3 strata: those having at least one out-migrant and received at least one remittance from the out-migrant during last 365 days (stratum 1); remaining households with at least one other type of migrants including the temporary out-migrants, for employment purpose (stratum 2); and all remaining households (stratum 3). A sample of 10 households was selected with break-up for 2, 4 and 4 households respectively for strata 1, 2 and 3 was selected by SRSWOR. In the FSUs with segment formation, the sample allocation to each of the selected two segments was 5 each with break-up as 1, 2 and 2 respectively for strata 1, 2 and 3 and the sample households were selected by simple random sampling.

- **NSS 65th Round (July 2008–June 2009)**

[Domestic tourism, housing condition, slum.]

A comprehensive survey on domestic tourism was conducted in NSS 65th round as per the request of Ministry of Tourism, Government of India to fill up the data gaps required for preparation of tourism satellite account. In addition, the subjects of housing condition and slum formed part of this round. It may be noted that the survey on consumer expenditure that was being carried out on a thin sample basis during NSS 42nd round till 64th round was dropped from NSS 65th round. In this round also, the survey on slums was confined to only urban areas of the country as in NSS 58th round. Only the slums found in the randomly selected urban blocks were surveyed.

As regards the stratification of FSUs, a broadly similar approach as in NSS 58th round, discussed earlier, was adopted in this round also with the following changes: (i) no special strata of small/large villages were formed and (ii) the urban stratum of the lowest population size class of towns (comprising towns having population of less than 100,000) within each NSS region was divided into 2 strata of towns i.e., those having population less than 50,000 and the others with population of 50,000 to less than 100,000. Within each urban stratum, 2 sub-strata of blocks lying in slum area and the residual blocks were also formed as in 58th round.

Altogether 12,952 FSUs, namely villages and UFS blocks combined, were selected for survey in this round as the central sample. In the method of total sample allocation of FSUs of a State/UT between rural and urban areas, instead of double weight to urban given in 58th round, 1.5 weight was given in this round. However, the double weight given to sub-stratum 1 comprising UFS blocks lying in slum area, as in 58th round, continued in NSS 65th round as well. The method of sampling of FSUs was changed from PPSWR/SRSWOR to circular systematic sampling. Villages were arranged as per census arrangement and FSUs were selected by circular systematic sampling with probability proportional to population from all the rural strata. The towns within the urban stratum/sub-stratum were arranged in the ascending order of population and thereafter the sample FSUs i.e., UFS blocks were selected by circular systematic sampling with equal probability.

In the selection of HGs/SBs from large FSUs, there was some change as compared to 58th round. The HG/SB having maximum population was selected with probability '1' and one more was selected by simple random sampling.

A stratified random sample of 12 households was selected from each FSU for each of the enquiries on domestic tourism and housing condition. The household stratification and sample allocation for the respective subjects of domestic tourism (Schedule 21.1) and housing condition (Schedule 1.2) were as follows (Box 3.7):

The required number of sample HHs from each stratum was selected by SRSWOR.

Box 3.7 Stratification of households and sample allocation for the surveys on domestic tourism and housing condition in NSS sixty-fifth round

Stratum	Stratum composition	No. of sample HHs per FSU*
Schedule 21.1 (Domestic Tourism)		
1	HHs having pucca dwelling (rural)/MPCE $\geq$ B (urban) and having at least one member who performed at least one overnight trip during last 30 days	4
2	HHs not having pucca dwelling structure (rural)/HHs with MPCE < B (urban) and having at least one member who performed at least one overnight trip during last 30 days	2
3	Remaining HHs having pucca dwelling structure (rural)/MPCE $\geq$ B (urban) and having at least one member who performed at least one same-day trip during last 30 days	2
4	Remaining HHs not having pucca dwelling structure (rural)/HHs with MPCE < B (urban) and having at least one member who performed at least one same-day trip during last 30 days	2
5	All remaining HHs	2
Schedule 1.2 (Housing Condition)		
1	HHs having pucca dwelling structure (rural)/MPCE of top 30% of urban population i.e., MPCE $\geq$ B	4
2	HHs having semi-pucca dwelling structure (rural)/MPCE of middle 40% of urban population i.e., A < MPCE < B	4
3	Other HHs (rural)/HHs having MPCE of bottom 30% of urban population i.e., MPCE $\leq$ A	4

Note Values of 'A' and 'B' were determined from NSS 61st round data for each NSS region in such a way that the top 30% of the urban population had MPCE $\geq$ B and the bottom 30% of the urban population had MPCE $\leq$ A.

*Sample allocation for each stratum was half the above allocations for each HG/SB in case of large FSUs.

Sources MoSPI (2010a, 2010b)

- **NSS 68th Round (July 2011–June 2012)**

[Household consumer expenditure, employment and unemployment.]

NSS 68th round was the last quinquennial round/survey on employment–unemployment (E-U) and household consumer expenditure (HCE) which was introduced in NSS 27th round. Subsequently, the subject of E-U is being covered annually through the Periodic Labour Force Survey (PLFS) introduced during 2017–18.

The list of 2001 Census villages (henceforth the term 'village' will mean panchayat wards in case of Kerala) constituted the sampling frame for rural areas. In urban, the latest (2007–12) list of UFS blocks served as the sampling frame of FSUs. As regards the first-stage stratification, each district, separately for rural and urban areas, was treated as a stratum. It is of interest to mention that even for urban areas each district is being generally treated as a separate stratum in the NSS since its 61st round (except 65th round and the PLFS). Each million-plus city as per Population Census 2001, if any, within a district was considered as a stratum and the rest of the cities and towns of the district together formed another stratum.

If 'r' was the allocation of sample villages to a stratum (made multiples of 4 to allocate equal number of sample villages to each sub-round), r/4 sub-strata of almost equal size (i.e., total rural population) was formed after arranging the villages of the stratum in ascending order of population. Similarly, in the urban areas, if 'u' was the allocation of sample FSUs to a stratum (multiple of 4), u/4 sub-strata of more or less equal size (i.e., total number of households as per the UFS) was formed after arranging the towns in ascending order of number of households in the town as per UFS phase 2007–12 and then arranging the IV units of each town and the blocks within each IV unit in ascending order of their numbers. Finally, 4 sample FSUs were allocated to each stratum x sub-stratum in both rural and urban areas.

A total sample size of 12,784 FSUs (rural and urban combined) was allocated for the central sample at all-India level. This was allocated to the States/UTs in proportion to their population as per Census 2001 (keeping in view the number of field investigators) subject to a minimum sample allocation to each State/UT. State/UT-level sample size was allocated between rural and urban in proportion to population with double weight to urban with the condition that the urban allocation for bigger/more urbanized States like Maharashtra, Tamil Nadu, etc. did not exceed the allocation for the rural areas. A minimum of 16 FSUs (8 each for rural and urban) was allocated to each State/UT. The State/UT level rural/urban sample size was further allocated to different strata/sub-strata in proportion to the population as per Census 2001. Allocations at stratum level were adjusted to multiples of 4 with a minimum sample size of 4. Allocation for each sub-stratum was 4. The allocated number of sample FSUs from each sub-stratum was selected by PPSWR in rural areas (with village population as per census 2001 as the size measure) and SRSWOR in urban. Both rural and urban samples were drawn in the form of 2 independent sub-samples with equal sample size allocated to each of the 4 sub-rounds of 3 months each. The FSU was surveyed during the sub-round to which it was allocated. The total number of surveyed FSUs at the all-India level for the central sample was 7,469 villages and 5,268 UFS blocks.

Large FUSUs with present population of 1,200 or more (600 or more in case of certain hilly areas) were divided into a minimum of 3 HGs/SBs by equalizing population. The one having maximum population was selected with certainty and one among the remaining HGs/SBs was selected by SRS for listing purposes.

So far as survey on HCE is concerned, two schedule types of HCE were designed for data collection i.e., Type I Schedule and Type 2 Schedule.[33] Listed households in each FSU/HG/SB were stratified into 3 strata as follows (Box 3.8) and a sample of 8 HHs was selected from each FSU independently for collecting data on E-U (through Schedule 10) and HCE (through Schedule 1.0) as per the stratification and sample allocation as follows:

From each SSS, the sample households were selected by SRSWOR independently for each schedule.

- **NSS 73rd Round (July 2015–June 2016)**

[Unincorporated non-agricultural enterprises belonging to manufacturing, trade and other services (excluding construction).]

By utilizing the list of 2011 census villages and census enumeration blocks (CEBs) with information on number of enterprises and workers as per the sixth Economic Census (EC)—conducted during 2013–14—as the sampling frame of villages/blocks, the first comprehensive survey of unincorporated non-agricultural enterprises engaged in manufacturing, trade and other services was conducted during NSS 73rd round. The survey covered the entire country. Within the manufacturing sector, the factories registered under Factories Act, 1948 which were covered in the Annual Survey of Industries (ASI) were left out of the coverage of this round to avoid duplication. As usual, the entire survey period was divided into 4 sub-rounds of 3 months each. Equal number of villages/blocks was allocated to each sub-round with fieldwork in any sample FSU was carried out during the sub-round to which it was allotted.

As stated above the list of villages and CEBs as per 2011 Census with information on number of enterprises/workers by industry as per the sixth EC served as the frame of FSUs (villages/CEBs) wherever the EC data was available. Otherwise, the 2011 Population Census villages and UFS blocks served as the sampling frame of villages/

[33] Type-1 Schedule had a uniform reference period of last 30 days for all food and non-food items along with 365-day reference period for clothing, bedding, footwear, education, medical (institutional), and durable goods. On the other hand, Type-2 Schedule had varying reference periods of (i) last 365 days for clothing, bedding, footwear, education, medical (institutional), and durable goods; (ii) last 7 days for edible oil; egg, fish & meat; vegetables, fruits, spices, beverages and processed foods; pan, tobacco & intoxicants; and last 30 days for all other food, fuel and light, miscellaneous goods and services including non-institutional medical; rents and taxes. In the NSS 68th round, two sets of average MPCE (monthly per capita expenditure) figures were arrived at using the data of Type-1 Schedule: one using the data of 'last 30 days' for all items (known as MPCE: URP i.e., Uniform Reference Period of last 30 days) and the other using 'last 365 days' data for clothing, bedding, footwear, education, medical (institutional), and durable goods and 'last 30 days' data for the rest of the items (described as MPCE: MRP i.e., Mixed Reference Period). The estimate of average MPCE based on Type-2 Schedule was called as MPCE: MMRP i.e., Modified Mixed Reference Period.

Box 3.8 Stratification and sample allocation of households in NSS sixty-eighth round

Second-stage stratum (SSS)	Composition of SSS	Number of households selected for Sch. 10/ Sch.1.0: Type 1/Type 2	
		FSU without HG/SB formation	FSU with HG/SB formation (for each HG/SB)
Rural			
1	Relatively affluent HHs*	2	1
2	Of the remaining, HHs having principal earning from non-agricultural activity	4	2
3	Remaining HHs	2	1
Urban			
1	HHs having MPCE of top 10% of urban population (MPCE > B)**	2	1
2	HHs having MPCE of middle 60% of urban population (A ≤ MPCE ≤ B)**	4	2
3	HHs having MPCE of bottom 30% of urban population (MPCE < A)**	2	1

Notes: * Affluent HHs were identified on the basis of a number of criteria such as possession of certain durables or assets, ownership of large business, highly remunerative profession, high salaried income, ownership of spacious pucca house in good condition, etc. If there were more than 10 such households, the 10 most affluent HHs were identified as the 'relatively affluent households' of the village.
** MPCE cut-off points 'A' and 'B' (in Rs.) were determined for each NSS region in such a way that the top 10% of the population had MPCE more than 'B' and bottom 30% of the population had MPCE less than 'A', as estimated by the NSS 66th round survey.
Sources MoSPI (2014a, 2014b)

blocks. For drawing sample FSUs (both rural and urban) of 'semi-round 1' i.e., sub-rounds 1 and 2 combined, EC data was available and could be used only for 21 States/ UTs.[34] However, for sample selection of 'semi-round 2' (sub-rounds 3 and 4), EC data was used for all the States/UTs.

For both rural and urban areas, each district was treated as a separate stratum. However, if the district contained any city with a population of 1.5 million or more, each such city formed a separate stratum.

[34] See MoSPI (2018) for the names of these States/UTs.

In the rural areas where EC data was used for sampling of FSUs, each stratum was divided into 3 sub-strata: villages having at least 3 establishments (i.e., with at least one hired worker) under certain pre-specified rare industries under survey coverage[35] (sub-stratum 1); out of the remaining villages, those having at least 8 establishments under coverage in the manufacturing and services sectors (sub-stratum 2); and the remaining villages of the stratum (sub-stratum 3). In the rural areas of Kerala, same procedure of sub-stratification was followed with the only change that CEB was the FSU instead of village. For the rural areas of the States/UTs for which EC data could not be used, a stratum was divided into r/4 sub-strata (with 'r' denoting the allocation of sample FSUs to a stratum—adjusted to multiple of 4) after arranging the villages according to number of non-agricultural workers in the village as per Primary Census Abstract (PCA), 2011 such that each sub-stratum comprised a group of arranged villages in a manner that finally each such sub-stratum had more or less equal size in terms of total number of non-agricultural workers as per the PCA.

As regards the urban areas, for each city with population of 1.5 million or above forming a separate stratum, EC data was used for sampling of FSUs i.e., CEBs. Each such stratum was divided into 14 sub-strata of CEBs (mutually exclusive and exhaustive) with each CEB within a sub-stratum containing at least one establishment in some rare industries, namely, CEBs with at least one establishment in storage & warehousing (sub-stratum 1); remaining CEBs with at least one establishment in accommodation, event catering and other food service activities (sub-stratum 2); remaining CEBs with one or more establishment in real estate, legal, accounting, management consultancy, professional, scientific and technical etc. (sub-stratum 3); ....and so on.[36] For the other urban strata (excluding cities with population of 1.5 million or above) of the States/UTs where EC data was used, each stratum was divided into 3 sub-strata exactly as per the same procedure adopted for rural strata with EC frame used. For urban areas of the States/UTs where EC data could not be used and instead UFS list of blocks was used as the frame, each stratum was divided into 2 sub-strata, namely, UFS blocks identified as Bazaar area/Industrial area/Hospital area/ Slum area (sub-stratum 1); and remaining UFS blocks of the stratum (sub-stratum 2).

Total sample size of FSUs (central sample) in this round was fixed at 16,348 and the same was allocated over the States/UTs in proportion to total number of non-agricultural workers in the industries under coverage as per the fifth EC. State/UT wise total sample size was further allocated to rural and urban areas in proportion to number of non-agricultural workers as per sixth EC (for States/UTs where EC frame was used) or as per PCA, 2011 (for the other States/UTs). A minimum of 8 FSUs (minimum 4 each for rural and urban separately) were allocated to each State/ UT. A similar method was used to allocate rural/urban sample size of a State/UT over the strata. Stratum level allocations were adjusted to multiples of 4. Stratum level allocations were further allocated over sub-strata in proportion to (i) number of

[35] See MoSPI (2018) for the names of the industries.

[36] For details, see MoSPI (2018).

non-agricultural workers as per sixth EC if EC frame was used; (ii) number of non-agricultural workers as per PCA, 2011 when Population Census frame was used; or (iii) number of UFS blocks with double weight to sub-stratum 1 when UFS frame was used. Efforts were made to allocate a minimum of 4 sample FSUs to a sub-stratum so that at least one FSU in each sub-stratum could be covered during each sub-round to tackle the seasonality issues.

Sample FSUs were drawn independently for each of the two semi-rounds. From each sub-stratum (rural/urban) where EC frame was used and also from each sub-stratum of rural areas with PCA, 2011 list used, sample FSUs were drawn by PPSWR with total number of non-agricultural workers as per the sixth EC or Population Census, 2011 (as applicable) as the size measure. For urban strata with UFS frame used, sample FSUs were drawn by SRSWOR. Both rural and urban samples were drawn in the form of two independent sub-samples for each semi-round (i.e., phase of six months) with equal number of samples allocated among the two sub-rounds of each phase.

In each FSU, all the large enterprises (under coverage) with at least 20 workers formed 'segment 9' and they were surveyed without any sampling. Thereafter, the large FSUs with approximate present population of 1,200 or more and/or number total number of non-agricultural enterprises as 120 or more was divided into a minimum of 4 HGs/SBs by more or less equalizing population/number of enterprises.[37] Out of all such HGs/SBs, the one having maximum number of enterprises under coverage was selected with certainty (identified as 'segment 1'). From the remaining, 2 HGs/SBs were selected by SRSWOR and merged for listing purposes (identified as 'segment 2'). Smaller FSUs without HG/SB formation were identified as segment 1.

All the eligible listed enterprises were stratified into 16 second-stage strata (SSS)—3 strata of 'establishments' (i.e., enterprises with at least one hired worker) in the manufacturing sector, 3 strata of establishments in the trade sector, 7 strata of establishments in services sector (other than trade), and 3 strata of own account enterprises (i.e., without any hired worker) with one each for manufacturing, trade and services sector. The stratification of establishments within each of manufacturing/ trade /services sector was based on different broad activities of the establishments within the respective sectors (see MoSPI (2018) for further details). A sample size of 2 enterprises/establishments was allocated to each of the 16 s-stage strata. In large FSUs, 1 enterprise/establishment was allocated to each of the 16 strata for each of segments 1 and 2. Thus, a total sample of 32 enterprises per FSU was targeted for data collection. There were rules for compensation of shortfall in required number of enterprises in any stratum from the other stratum/strata. From each stratum (SSS), the sample enterprises were selected by SRS (SRSWOR in case allocation was 2 or more).

---

[37] See MoSPI (2018) for further details.

- **NSS 74th Round (July 2016–June 2017)**

[Services sector enterprises (List frame-based survey of 'bigger' enterprises).]

The coverage of the NSS rounds on enterprise surveys conducted in the past was confined mainly to unorganized/unincorporated enterprises. These surveys adopted a stratified multi-stage design where area frames of villages/urban blocks were considered for the selection of villages/blocks as the FSUs. There was no comprehensive dedicated survey of bigger enterprises in the services sector as in the case of manufacturing sector where large units are covered annually through the Annual Survey of Industries (ASI). A need was felt over a considerable period of time to put in place an annual survey of larger units in the services sector through a list frame-based survey of such units on the lines of ASI. NSS 74th round was a modest beginning in this venture.

Within the services sector, the types of enterprises included proprietary/partnership enterprises, limited liability companies, non-government companies, cooperative Societies, non-profit Institutions and trusts. The survey excluded from its coverage government enterprises/PSUs, private money Lenders and self-help groups (SHG). Air transport and financial and insurance sector were also kept out of coverage of 74th round.

The sampling frame for the survey was developed by considering three frames namely (i) sixth EC frame of establishments with 10 or more workers, (ii) business register (BR) of establishments with 10 or more workers available for 12 States,[38] and (iii) the list of active private non-financial companies of 2013–14 as per the Ministry of Corporate Affairs (MCA) frame of companies obtained by the NSSO from the National Accounts Division (NAD) of CSO.[39] The first three months (Phase—I) of the survey period were utilized to verify/update the frames of EC and BR while the next nine months (Phase—II) were utilized to carry out the actual survey in a sample of units drawn from the finally verified frame of EC/BR and the list of companies. The detailed schedule was also canvassed in some big construction enterprises found eligible for survey from EC and BR frames as also from MCA frame for use of NAD.

Although it was planned to verify all the establishments in list frame of EC and BR having 10 or more workers, due to the constraint on resources, the said verification could be carried out in respect of establishments with 16 or more workers only. In addition, for each State × Broad Activity[40] combination which had only less than 10 establishments (irrespective of number of workers), this verification was possible. This finally verified list from EC and BR frames—being a truncated list in the sense that a large number of establishments with 10 or more workers were excluded—along with the MCA frame of companies constituted the ultimate sampling frame

[38] Andhra Pradesh, Arunachal Pradesh, Himachal Pradesh, Lakshadweep, Manipur, Nagaland, Punjab, Rajasthan, Tamil Nadu, Telangana, Uttarakhand and Uttar Pradesh.

[39] MoSPI (2019)

[40] Broad activities were in terms of major industries i.e., National Industrial Classification (NIC, 2008) Sections like Section G (Wholesale and retail trade, repair of motor vehicles and motorcycles), Section H (Transportation and storage), Section I (Accommodation and food service activities), and so on.

Box 3.9 Total number of units in the fame and sample size, NSS seventy-fourth round

Type of frame	Frame size* (Total number of units)	Sample size
Sixth EC	56,664	25,257
BR	6,426	2,749
MCA	349,500	35,453

Note * Number of verified units as per EC/BR frame).
Source MoSPI (2019)

of NSS 74th round. The total number of establishments taken up for verification at the all-India level was 134,982 (108,622 based on EC frame and 26,360 from BR frame). After verification, only 63,090 establishments (EC: 56,664; BR: 6,426) establishments (including those in the construction sector) were found to be eligible for the survey. Box 3.9 shows the size of finally updated frame and sample size adopted for the survey by type of sampling frame.

The eligible enterprises in the ultimate frame of EC/BR were divided into two sectors i.e., complete enumeration and sample sectors. The former comprised (i) establishments with 55 or more workers; and (ii) State x Broad Activity strata with less than 6 establishments. For the establishments in the sample sector, each State x Broad Activity combination (with at least 6 establishments) was treated as a separate stratum with proportionate allocation of samples as per total number of workers in the respective strata. The required number of units from each stratum in the sample sector was selected by circular systematic sampling after arranging them by number of workers.

Like the units in the frame of EC/BR, the MCA list of units/companies was also divided into complete enumeration sector and sample sector. The complete enumeration sector included (i) all companies reporting annual revenue of Rs. 20 Crore or more; (ii) all companies of the States with less than 100 Companies; and (iii) all companies belonging to a State × Industry group[41] (strata) having less than 5 units. The remaining companies i.e., strata having 5 or more units/companies formed the sample sector. Each stratum in the sample sector was sub-stratified into 3 sub-strata as follows: companies with annual revenue of Rs. 1 Crore or less (sub-stratum 1); companies with annual revenue between Rs. 1 Crore and Rs. 10 Crore (sub-stratum 2); and all remaining companies with annual revenue greater than Rs. 10 Crore but less than Rs. 20 Crore (sub-stratum 3). Finally, the required number of companies from each sub-stratum was selected by circular systematic sampling after arranging the companies as per their annual revenue.[42]

[41] For the actual number of industry groups formed and their composition, refer to MoSPI (2019): Statement 2 in Appendix B.

[42] For more details about the survey methodology of NSS 74th round and the survey findings, readers may refer to MoSPI (2019).

3.5 The Final/Fourth Phase (Since 2017–18)

- **Periodic Labour Force Survey (July 2017–June 2018) and Onwards till December 2024**

Keeping in view the growing demand for employment–unemployment (E-U) statistics at more frequent intervals, the Periodic Labour Force Survey (PLFS) was introduced during 2017–18 with mainly two main objectives, namely, to provide (i) annual estimates of important E-U parameters as per both 'usual activity status'[43] and 'current weekly activity status'[44] approaches for rural as well as urban areas, and (ii) quarterly estimates of key labour force indicators i.e., worker population ratio (WPR), labour force participation rate (LFPR), and unemployment rate (UR) as per current weekly activity status for urban areas. In order to provide quarterly estimates, the urban sample size of FSUs for each quarter was kept more or less similar to (little higher than) the annual urban sample size of the last (NSS 68th round) quinquennial survey on the subject. In the PLFS, at the all-India level, nearly 7,000 villages and a little more than 5,700 UFS blocks are generally surveyed every year. As against this, 7,469 villages and 5,268 UFS blocks were surveyed during NSS 68th round. While PLFS annual estimates of E-U are available from the year 2017–18, the quarterly estimates are available since the quarter of April–June 2018.

One distinguishing feature of the urban component of the PLFS is the adoption of rotational panel sampling design where each FSU (i.e., UFS block) and each sample household is visited in 4 consecutive quarters with 'first visit' schedule canvassed in the households when they are visited for the first time and 'revisit' schedule canvassed in the subsequent three visits/quarters. To facilitate the implementation of rotational panel design, sample FSUs from each urban stratum are drawn in the form of 4 sets/panels of FSUs with each set/panel having equal number of FSUs. Each panel remains in the sample for 4 consecutive quarters of the survey (PLFS) and then the panel disappears. The disappeared panel in a given quarter is replaced with a new panel. Between any two consecutive quarters, three panels are common and that way nearly 75% of the sample households are common between any two consecutive quarters in the PLFS. The rotational panel scheme has been introduced since such a scheme is likely to provide better estimates of change parameters between two quarters. As an illustration, the rotational panel scheme (urban areas) as well as the sampling scheme for rural areas for the two-year period (July 2021–June 2023) is provided in Box 3.10 (MoSPI, 2023).

As usual, the PLFS adopts a stratified multi-stage design with villages/UFS blocks as the FSUs and households as the ultimate stage units with selection of HGs/SBs in the large FSUs. However, as compared with NSS 68th round survey, there are

[43] The activity status on which a person spends relatively long time (major time criterion) during the 365 days preceding the date of survey is considered the usual principal activity status of the person. This is a concept adopted in the previous employment–unemployment surveys as well.

[44] The current weekly activity status of a person is the activity status for a person during a reference period of 7 days preceding the date of survey.

Box 3.10 Rotational plan scheme for urban areas and sampling scheme for rural areas for the PLFS: 2021–22 & 2022–23

	Panels for the third two-year period							
	During the four quarters of the fifth year				During the four quarters of the sixth year			
	Qtr1	Qtr2	Qtr3	Qtr4	Qtr1	Qtr2	Qtr3	Qtr4
Urban	P_{31}*	P_{31}	P_{31}	P_{31}	P_{35}*	P_{35}	P_{35}	P_{35}
	P_{28}	P_{32}*	P_{32}	P_{32}	P_{32}	P_{36}*	P_{36}	P_{36}
	P_{27}	P_{28}	P_{33}*	P_{33}	P_{33}	P_{33}	P_{37}*	P_{37}
	P_{26}	P_{27}	P_{28}	P_{34}*	P_{34}	P_{34}	P_{34}	P_{38}*
Rural	R_{31}*	R_{32}*	R_{33}*	R_{34}*	R_{35}*	R_{36}*	R_{37}*	R_{38}*

Note *Visit 1 schedule canvassed.
R_{ij}* indicates rural samples covered in j-th quarter of i-th two-year period.
P_{ij} indicates the panel belonging to j-th quarter of the i-th two-year period of rotation.
Source MoSPI (2023)

some changes in the method of stratification and sampling of FSUs as well as in the stratification of households in the PLFS. The most significant ones are as follows:

- While NSS 68th round treated each district as a basic stratum in both rural and urban areas, PLFS considers each NSS region instead as the basic stratum (rural & urban both).
- Each city with a population of 1 million or above as per census 2001 constituted a separate stratum and all other towns of each district together formed a stratum in NSS 68th round. However, in the PLFS, each city with a population of 1.5 million or more as per census 2011 forms a separate stratum. All other cities and towns within each NSS region are divided into 3 strata as per their census 2011 population: all towns with population less than 50,000 (stratum 1); all towns with population 50,000 or more but less than 300,000 (stratum 2); and all towns with population 300,000 or more but less than 1.5 million (stratum 3). Thus, generation of district-wise estimates based on the PLFS is not feasible although NSS 68th round had that provision.
- Urban sample FSUs in NSS 68th round were selected by SRSWOR while they are selected by PPSWR in the PLFS with number of households in the FSU (i.e., UFS block) as the size measure.
- Stratification of households in NSS 68th round was primarily based on the criterion of relative living standards of the listed households (judged in terms of level of MPCE of households in urban and the level of relative affluence in case of rural areas) as described earlier. On the contrary, the PLFS considers the number of members in the household (say, M) having level of general education as secondary (10th standard) or above as the stratification variable to form 3 s-stage strata (SSS) in rural areas i.e., M = 2 or more (SSS 1), M = 1 (SSS 2) and M = 0 (SSS 3); and 4 SSS in urban areas: M = 3 or more (SSS 1), M = 2 (SSS 2), M = 1 (SSS 3), and M = 0 (SSS 4).

Allotted number of sample households per FSU is same (i.e., 8) in both NSS 68th round and PLFS. At the all-India level, total number of households surveyed in PLFS: 2022–23 was 55,844 in rural and 45,811 in urban as against 59,700 households and 42,024 households surveyed respectively in rural India and urban India during NSS 68th round.

- **Time Use Survey (January 2019–December 2019)**

A comprehensive survey on time spent by the people of India in various paid and unpaid activities covering the entire country was conducted for the first time during January–December 2019. Earlier, the MoSPI conducted a Time Use Survey (TUS) during July 1988–June 1999 on a much smaller scale (i.e., pilot basis) covering 18,591 households from six States of Gujarat, Haryana, Madhya Pradesh, Meghalaya, Odisha and Tamil Nadu. This pilot TUS adopted a stratified three-stage design with districts, villages/urban blocks, and households as the units of sampling in different stages. Within each State, districts were stratified into 4 strata by jointly considering population density and proportion of tribal population in the respective districts. Altogether, 52 districts out of 147 districts considering all these States were covered. A sample of 1,066 villages and 488 urban blocks was selected for the survey with an allocated sample of 12 households per village/block.

TUS (2019) covered the entire country except certain villages of Andaman & Nicobar Islands which were not accessible. The household questionnaire for the survey was designed to collect information on activity particulars of each household member of age 6 years and above, with a reference period of 24 h starting from 4 AM on the day before the date of interview to 4 AM on the day of the interview. The reference period was split into 48 time slots, each of duration of 30 min to record the activity particulars of the household members.

TUS (2019) had targeted to cover 10,004 FSUs (rural: 5,992 and urban: 4,012) at the all-India level. The survey adopted a stratified multi-stage design. List of census 2011 villages/latest list of UFS blocks served as the frame of FSUs with the change that the large villages with census 2011 population of 1,000 or more and large UFS blocks with 250 or more households were artificially divided into a certain number of segments called 'sub-units' (SUs) and these SUs were treated as the FSUs in respect of such larger villages/blocks for sampling purposes.[45] The TUS considered each NSS region as the basic stratum for both rural and urban areas instead of district being generally treated as a stratum in most of the NSS rounds. However, in the rural areas a special stratum was formed at the all-India level which comprised all the uninhabited villages of the country as per Census 2011 before the rural strata of inhabited villages as per the census were formed within each NSS region of a State/UT. All the villages of a rural stratum were further stratified into 3 groups: villages (Panchayat wards in case of Kerala) with Census 2011 population less than 250 (Group 1); villages (Panchayat wards for Kerala) with Census 2011 population more than or equal to 250 but less than 500 (Group 2); and the remaining villages

[45] This concept of SU formation in large villages/blocks was introduced in the 76th round of the NSS.

(panchayat wards for Kerala) forming Group 3. For urban areas, all towns within each NSS region were stratified into 3 population size classes of towns (i.e., less than 50,000; 50,000 to less than 300,000; and 300,000 to less than 1.5 million) apart from considering each city with a population of 1.5 million or more as a separate stratum. Both rural and urban FSUs within each stratum were further arranged in ascending order of village population/number of households in the UFS blocks to form a few sub-strata of roughly equal size (i.e., equal population in case of rural and equal number of households in case of urban) with a sample allocation of 4 FSUs to each sub-stratum. Finally, the sample FSUs from each sub-stratum were selected by SRSWOR. In each sample FSU, all the households were listed and a sample of 14 households was selected by SRSWOR for canvassing the household schedule in 2 households for each day of the week. It may be mentioned that if the approximate present population of the selected FSU (i.e., village/urban block/SU) exceeded 1,500 (750 in case of certain hilly areas[46]), the listing exercise was confined to a randomly selected segment of the FSU. For this purpose, the FSU was segmented into a minimum number of 2 segments by equalizing population. A pre-specified rule was provided to the field investigators to decide on the total number of segments required to be formed in such FSUs.

The second comprehensive TUS following the similar methodology was conducted during January–December 2024 with a total all-India sample allocation of 10,020 FSUs (rural: 5,992 and urban: 4,028).

- **Annual Survey of Unincorporated Sector Enterprises (April 2021–March 2022)**

Keeping in view the need for more frequent (annual) data pertaining to unincorporated enterprises, the 'first full-fledged' Annual Survey of Unincorporated Sector Enterprises (ASUSE) was conducted by the NSSO during April 2021–March 22 with broadly a similar coverage of unincorporated non-agricultural establishments in manufacturing, trade and other services sector as in NSS 73rd round (July 2015–June 2016). As the name suggests, the survey is proposed to be conducted every year.[47] The subsequently conducted surveys include ASUSE: 2022–23 (October 2022–September 2023), ASUSE: 2023–24 (October 2023–September 2024) and ASUSE: 2025 (January–December, 2025).

The fieldwork of ASUSE: 2021–22 was carried out in the form of four sub-rounds with each sub-round spread over a duration of three months. The general phenomenon of covering roughly equal number of sample size of FSUs in each NSS sub-round could not be implemented in ASUSE: 2021–22 since the second wave of the pandemic had stalled the initial phase of the survey period. The list of census

[46] These included rural areas of (i) Himachal Pradesh, (ii) Sikkim, (iii) Andaman & Nicobar Islands, (iv) Uttarakhand (except four districts Dehradun, Nainital, Hardwar and Udham Singh Nagar), (v) Punch, Rajouri, Udhampur, Reasi, Doda, Kishtwar, Ramban of Jammu, (vi) Ladakh and (vii) Idukki district of Kerala.

[47] In fact, the NSSO planned to conduct the first survey (i.e., ASUSE) during April 2020–March 2021. However, the survey could not be launched during the year 2020–21 due to pandemic and it was initiated from April 2021.

2011 villages with count of number of establishments/workers as per the sixth EC served as the frame of FSUs in rural areas except for Kerala where list of Panchayat wards served as the frame of FSUs. For urban areas, the latest list of UFS blocks was used as the sampling frame of FSUs. However, in NSS 73rd round, so far as urban areas are concerned, the 2011 census enumeration blocks (CEBs) with information on number of establishments/workers at the level of CEB as per the sixth EC mostly served as the sampling frame of FSUs.

ASUSE: 2021–22 treated NSS region as the basic stratum for both rural and urban areas instead of district as the stratum in NSS 73rd round. However, for rural areas, all the villages with no workers under the survey coverage considering the entire country together formed a special stratum before the basic strata at the NSS region level were formed. In urban areas, cities and towns within each NSS region were further stratified according to their population (P) as per census 2011 as follows: all towns with $P < 50{,}000$ (stratum 1); all towns with $P = 50{,}000$ to less than 300,000 (stratum 2); all cities/towns with $P = 300{,}000$ or more but less than 1 million (stratum 3); and each city with $P = 1$ million or more (stratum 4, 5, 6, …).

Each basic rural stratum was divided into 4 sub-strata/'groups' namely, all villages with at least one establishment appearing in the EC list frame used for service sector enterprises in NSS 74th round (Group 1)[48]; out of the remaining, all villages with less than or equal to 10 workers under survey coverage (Group 2); out of the remaining, all villages with 500 or more workers under survey coverage (Group 3); and all the remaining villages of the stratum (Group 4). Each urban stratum was sub-stratified into 2 sub-strata: UFS blocks identified as bazaar area/industrial area/hospital area/ Slum area (sub-stratum 1); and all the remaining UFS blocks of the stratum (sub-stratum 2).

An all-India sample size of 16,356 FSUs (8,468 in rural and 7,888 in urban) was considered for the survey. This sample size was allocated to States/UTs x strata x sub-strata in proportion to number of non-agricultural workers under coverage of the survey as per the sixth EC with a minimum allocation of 8 FSUs to both rural and urban for each State/UT. Stratum/sub-stratum level allocations were adjusted to multiples of 4 and the sample FSUs from each stratum/sub-stratum were selected by SRSWOR.

Within each sample FSU, Segment 9 was carved out which comprised all 'big' establishments i.e., those with 10 or more workers located within the boundaries of the entire sample FSU. Out of all such establishments, those under the survey coverage and also operated for at least one day during the preceding 365 days were surveyed on a complete enumeration basis.

After carving out segment 9, the entire FSU was considered for HG/SB formation as per the information on approximate present population (P) and approximate present number of non-agricultural establishments (E) in the whole FSU. The

[48] EC list frame comprises all the establishments having 10 or more workers as per the sixth EC. For the formation of Group 1, the establishments which could not be surveyed during the NSS 74th round were ignored.

number (D) of HGs/SBs (minimum 4) formed (Box 3.11) was guided by the following criterion:

Thus, if both P was less than 1,200 and E was less than 120, the entire FSU was considered for the listing of establishments. The above class-limits of population (and not the number of establishments) was reduced to 50% for certain hilly areas namely, rural areas of (i) Himachal Pradesh, (ii) Sikkim, (iii) Andaman & Nicobar Islands, (iv) Uttarakhand (except four districts Dehradun, Nainital, Hardwar and Udham Singh Nagar), (v) Punch, Rajouri, Udhampur, Reasi, Doda, Kishtwar, Ramban, Ladakh region (Leh and Kargil districts) and (vi) Idukki district of Kerala.

Out of the total number of HGs/SBs formed (i.e., D), the one having maximum number of establishments with hired workers under the survey coverage was selected with certainty (called segment 1) and from among the remaining (D-1) HGs/SBs, two more HGs/SBs were selected by SRSWOR and both of them were combined together, called segment 2, for a combined listing of establishments. Listing and sampling was done independently for segment 1 and segment 2.

In each of segments 1, 2 and 9, all those establishments under the survey coverage which operated for at least 30 days (15 days for seasonal establishments and self-help groups) during the reference period of last 365 days qualified for the survey.

The listed establishments (eligible ones) of segments 1 and 2 in an FSU were stratified into a maximum of 18 s-stage strata (SSS) by jointly considering (i) type of ownership of the establishment, (ii) major industry to which the establishment belonged, and (iii) whether the established had at least one hired worker or not. The details of stratification and the sample allocation for each SSS (with a maximum of 46 establishments considering all the 18 SSS) were as follows (see Box 3.12), in addition to the survey of all the eligible establishments from segment 9:

The required number of sample establishment from each SSS was selected by SRSWOR.

Box 3.11 Criterion for number of HG/SB formation in annual survey of unincorporated sector enterprises (April 2021–March 2022)

Value of P	Number of HGs/SBs as per the population criterion*	Value of E	Number of HGs/SBs as per the establishment criterion*
Less than 1,200	1	Less than 120	1
1,200–1,500	4	120–159	4
1,600–1,999	5	160–199	5
2,000–2,399	6	200–239	6
And so on	…	And so on	…

Note * D was taken as the higher of the two values as per the population and establishment criteria.
Source MoSPI (2024b)

Box 3.12 Composition of second-stage strata and sample allocation in annual survey of unincorporated sector enterprises (April 2021–March 2022)

SSS number	Composition of the SSS	Sample allocation*
1	Type of ownership being Society/Trust/Association/Club/Body of individuals, etc. (any industry/NIC code)	2
2	Type of ownership as Cooperatives (any industry/NIC code)	2
3	Type of ownership as Self-Help Groups	2
Establishments with type of ownership being other than the above:		
A. Those having hired workers		
(i) Manufacturing establishments		
4	Cotton ginning, cleaning and bailing, coke and refined petroleum products, pharmaceuticals, medicinal chemical and botanical product, computer, electronic and optical products, other transport equipment, electric power generation etc. (NIC 2008 Codes: 01,632, 19, 21, 26, 30, 35,103, 35,105, 35,106, 35,107, 35,109)	2
5	Food products and beverages (NIC Codes: 10, 11)	2
6	Textiles, wearing apparel, leather and related products, wood and wood products, paper and paper products, printing, furniture etc. (NIC Codes: 13, 14, 15, 16, 17, 18, 31)	2
7	Tobacco, chemical and chemical products, rubber and plastic products, other non-metallic mineral products, basic metals, fabricated metal product, electrical equipment, machinery and equipment n.e.c, motor vehicles, trailers and semi-trailers, 'other manufacturing activities, repair and installation of machinery and equipment (NIC Codes: 12, 20, 22, 23, 24, 25, 27, 28, 29, 32, 33)	2
(ii) Trading establishments		
8	Wholesale and retail trade and repair of motor vehicles and motorcycles, Other wholesale trade and Other retail trade (NIC Codes: 45, 46, 47)	4
(iii) Other services establishments		
9	Food and beverage service activities (NIC Code: 56)	2
10	Education (NIC Code: 85)	2
11	Human health activities, residential care activities (NIC Codes: 86, 87)	2
12	Other services activities	4
B. Those without any hired workers (i.e., Own Account Establishments)		
13	Manufacturing activities	4
14	Trading activities	4
15	Food and beverage service activities	2
16	Education	2
17	Human health activities, residential care activities	2

(continued)

Box 3.12 (continued)

SSS number	Composition of the SSS	Sample allocation*
18	Other services activities	4
All	–	46

Note *Sample allocation was the half for each of segment 1 and segment 2 for the FSUs with HG/SB formation.
Source MoSPI (2024b)

- **NSS 79th Round (July 2022–June 2023)**

[Comprehensive annual modular survey (CAMS) for indicators related to education, out-of-pocket medical expenditure, use of mobile & internet, financial inclusion, ICT skills, possession of assets, drinking water, sanitation, energy use, birth registration, access to transport facilities etc.) including survey on AYUSH.]

NSS 79th round covered a diverse set of subjects with an aim to generate indicators related to education, out-of-pocket medical expenditure, ICT skills of the household members, possession of assets, etc. The survey, inter-alia, collected information related to drinking water, sanitation, energy use, birth registration, and access to transport facilities. In addition, the awareness about the AYUSH (Ayurveda, Yoga & Naturopathy, Unani, Siddha, Sowa Rigpa and Homoeopathy) and its use also formed part of this round.

The survey covered the entire country excluding some inaccessible villages of Andaman & Nicobar Islands. As usual, a stratified multi-stage design was adopted in the survey with the FSUs being villages/UFS blocks/sub-units in case of large villages/UFS blocks and the USUs being the households. A sample of 8,758 FSUs from rural areas and 6,540 FSUs from urban areas was surveyed in the central sample.[49] The targeted number of sample households per FSU was 20 for the CAMS and 12 for the enquiry on AYUSH.

For the selection of FSUs, the stratification was resorted to with each district being treated as the basic stratum in respect of both rural and urban areas. However, in rural areas, a special stratum was formed which comprised all uninhabited villages of the country. Further, each basic stratum (rural) was divided into 2 strata: the villages within a distance of 5 km from the district headquarter or from a city with more than 5 lakh population (stratum 1); and all the remaining villages (stratum 2). In case of urban, each million plus city as per Census 2011 formed a separate stratum and all the remaining towns within each district together formed another stratum.

Sub-strata were formed within each stratum for both rural and urban areas after arranging the villages/UFS blocks according to their size (i.e., population in case

[49] MoSPI (2024a)

of villages and number of households in respect of UFS blocks).[50] From each sub-stratum, the required number of sample FSUs was selected by SRSWOR.

Sample FSUs having approximate present population of 1,500 or more (750 or more in case of certain hilly areas in the rural) were sub-divided into a pre-specified number of sub-divisions which were formed with more or less equal population and one sub-division was selected at random for the listing exercise and preparation of the sampling frame of households.

The listed households were grouped into 2 s-stage strata (SSS) for both the enquiries on CAMS and AYUSH. In respect of CAMS, SSS 1 consisted of the households having at least one member (including deceased former member) hospitalized (including institutional childbirth) during last 365 days and the residual households formed SSS2. A total of 20 households were surveyed from each selected FSU out of which 5 households from SSS1 and 15 from SSS 2 were selected by SRSWOR.

So far as the survey on AYUSH is concerned, SSS 1 was composed of the households incurring expenditure for AYUSH treatment (as in-patient or out-patient for at least one member) during last 365 days or of the households with at least one member having knowledge about AYUSH systems and all the residual households formed SSS 2. Finally, a total of 12 households i.e., 8 households from SSS 1 and 4 households from SSS 2 were selected by SRSWOR from each of the sample FSUs.

- **Household Consumption Expenditure Survey (August 2022–July 2023)**

The subject of household consumption expenditure (HCE) has been covered in many NSS rounds since its first round. The subject was covered at more or less five-year intervals as part of the 'quinquennial series' of NSS rounds since 27th round (October 1972–September 1973) till 68th round (July 2011–June 2012). Subsequently, as part of NSS 75th round (July 2017–June 2018) also, the survey on HCE was conducted but the survey results were not released due to data quality issues. In between, the subject was covered on a 'thin sample' basis annually during NSS 42nd round till 64th round.

In view of the diverse uses of data on HCE, the most recent Household Consumption Expenditure Survey (HCES) was conducted during 2022-23 and the survey was repeated in the next year as the HCES: 2023-24.

HCES: 2022-23 covered entire geographical area of the country except a few villages of Andaman & Nicobar Islands which were inaccessible throughout the year. A stratified multi-stage design was adopted for the survey with census 2011 villages/latest UFS blocks/'sub-units' for the large villages/blocks as the FSUs and households as the USUs. The survey covered 155,014 rural households and 106,732 urban households spread over 8,723 rural FSUs and 6,115 urban FSUs. As against this, the last quinquennial survey (HCE) of NSS 68th round covered 7,469 villages and 5,268 UFS blocks with nearly 59,700 rural households and a little less than 42,000 urban households interviewed for each of 'Schedule Type 1' and 'Schedule Type 2'.

---

[50] For further details, see MoSPI (2024a).

As regards the survey methodology, there have been quite a few changes in the HCES: 2022-23 as compared with the same adopted in NSS 68th round. The major changes in the approach of data collection and in the sample design are as follows:

(i) The most significant change in the approach of data collection was in terms of the splitting of the traditional lengthy questionnaire into three parts comprising 'food items'; 'consumables and services items'; and 'durable goods' and these three parts were canvassed in each sample household through three consecutive monthly visits.

(ii) Regarding the sampling frame of FSUs, the large villages with census 2011 population of 1,000 or more (500 or more in case of hilly rural areas of certain States/UTs) and present likely projected population of 1,200 or more (600 or more as the case may be) were notionally divided into a few (minimum of 2) smaller units, called 'sub-units' (SU), of roughly equal population and each such SU was treated as the FSU for sampling purposes. All other smaller villages in terms of population were treated as the FSUs. SUs were also notionally formed in a similar manner in the UFS blocks with 250 or more households as per the latest UFS frame used for sampling.

(iii) The HCES treated each State/UT as a basic stratum in place of district as the stratum in 68th round.

(iv) All the villages of a State/UT (excluding the uninhabited villages, which for all States/UTs together formed a special stratum at the all-India level) were stratified into 2 strata: those located within a distance of 5 km from the district headquarter or from a city/town with more than 500,000 population (stratum 1); and other villages (stratum 2). Within each rural stratum (except the special stratum of all uninhabited villages for all-India), 3 groups of villages were formed as per the census 2011 population (P) of the villages: those with $P < 250$ (Group 1); $P = 250$ to 499 (Group 2); and $P = 500$ or more (Group 3). On the contrary, in 68th round, villages within a district were arranged according to population and a few sub-strata of equal size were formed.

(v) In the urban areas, all the UFS blocks of a State/UT were further stratified based on twin criteria of the population size class of town to which the UFS blocks belong and whether the blocks are identified as located in 'affluent area' or not as per the UFS frame. The composition of urban strata within a State/UT (Box 3.13) were as follows[51]:

If the sample size of FSUs allotted to a stratum was more than 20, two or more sub-strata were formed based on the number of FSUs within that stratum. As against the above, in NSS 68th round, all towns within a district (apart from million-plus city if any, each one of which formed a stratum) were arranged according to population to form a few sub-strata.

[51] MoSPI (2024c)

Box 3.13 Composition of strata of urban blocks in the household consumption expenditure survey (August 2022–July 2023)

Stratum number	Composition
11	'Affluent' UFS blocks of all towns with population less than 50,000
12	Remaining UFS blocks of all towns with population less than 50,000
21	'Affluent' UFS blocks of all towns with population 50,000 or more but less than 300,000
22	Remaining UFS blocks of all towns with population 50,000 or more but less than 300,000
31	'Affluent' UFS blocks of all towns with population 300,000 or more but less than 1 million
32	Remaining UFS blocks of all towns with population 300,000 or more but less than 1 million
Stratum 41, 42, 43…	Separate stratum for each of the million plus towns with odd and even numbers given for affluent and non-affluent blocks respectively

Source MoSPI (2024c)

(vi) From each sub-stratum, sample FSUs in the HCES for both rural and urban were selected by SRSWOR in the form of 10 independent panels with 2 independent sub-samples in each. On the other hand, sample FSUs from each sub-stratum in NSS 68th round were selected by PPSWR in case of rural and SRSWOR in urban, in the form of 2 independent sub-samples.

(vii) In the HCES, stratification of households within the FSU was based on the criterion of land possessed area in rural and the availability of four-wheeler car for non-commercial use in urban blocks—with 3 second-stage strata (SSS) of households formed in each FSU and a combined sample size of 18 households allocated to 3 strata. NSS 68th round stratified the households based on their relative affluence (rural)/MPCE level (urban) with a total sample size of 8 households per FSU.

As regards the questionnaire, for collection of data on food and non-food items, the traditionally canvassed long questionnaire comprising all these items—which was also canvassed in the previous survey during NSS 68th round—was split into 3 questionnaires, namely, (i) QUESTIONNAIRE—FDQ: FOOD ITEMS (with varying reference periods of 7 days/30 days for different items); (ii) QUESTIONNAIRE—CSQ: CONSUMABLES & SERVICES (having a varying reference periods of 7 days/30 days/365 days for different items); and (iii) QUESTIONNAIRE—DGQ: DURABLE ITEMS (with same reference period of 365 days for all the items). These three questionnaires were canvassed 'at random in a selected household in three consecutive monthly visits in a quarter'.[52] As per the convention followed, data for all the items in the three questionnaires were

[52] For more details about the contents of the respective questionnaires and the reference periods for different food and non-food items, the readers may refer to MoSPI (2024c).

collected through personal interview[53] by interviewing the appropriate respondents in the sample households. The change in the approach of data collection through split schedules and multiple visits to a household is likely to yield improved estimates of average MPCE and consequently the poverty rates based on HCES: 2022–23 and HCES: 2023–24 because of the reduction in respondents' fatigue. However, this has led to incomparability of these estimates with those as per the past series. Manna (2024) suggested a remedy to overcome this problem.

- **Periodic Labour Force Survey (January 2025–December 2025)**

PLFS 2025 is the revamped survey of the earlier PLFS series conducted during 2017–18 to December 2024 with the additional provisions to generate quarterly estimates of key labour force parameters (CWS approach) for rural areas also in addition to urban areas and to generate monthly estimates of key labour force parameters (all-India only) as per the CWS approach for both rural and urban areas. To take care of the same–particularly the generation of monthly estimates–the sample size of PLFS 2025 has been increased substantially to 12,504 FSUs in the rural areas and 10,188 FSUs in the urban areas. The corresponding sample allocations were much less in the previous series (7,016 rural FSUs and 5,784 urban FSUs in the PLFS: 2023-24). Further, the number of sample households per FSU in the PLFS 2025 has been increased to 12 as against 8 households per FSU in the earlier series. The rotational sampling scheme followed in the urban areas of the earlier series to produce quarterly estimates is extended to rural areas in PLFS 2025 with the change that there is a monthly rotation scheme for both rural and urban areas. Each month, starting with the fourth month i.e. April 2025 will have 4 panels of sample FSUs out of which one panel is the freshly introduced panel during the month for survey, while the other three panels are the panels of FSUs already introduced during the previous months. Each selected FSU/household will be visited four times in four consecutive months with a 75% matching between two consecutive months and a 50% matching between two consecutive quarters.[54]

References

Department of Statistics. (1974). *Some summary information on the survey of employment and unemployment, NSS 27th Round, October 1972—September 1973, Report Number 255/1*. Government of India.

Department of Statistics. (1978a). *Report on livestock number and products (some estimates on production of livestock products and related characteristics), NSS 30th round, July 1975—June 1976, NSS report number 281/7*. National Sample Survey Organisation, Government of India.

---

[53] A Computer-Assisted Personal Interview (CAPI) method was adopted in HCES: 2022–23 as against the Pen-and-Paper Personal Interview (PAPI) method followed in the previous surveys.

[54] For further details including the changes in the sample design, see MoSPI (2025c).

Department of Statistics. (1978b). *Tables with notes on survey of self-employed households in non-agricultural enterprises—Detailed results, State: Gujarat, Kerala, Tamil Nadu and West Bengal, NSS 29th round, July 1974—June 1975, Report number 280/1.* National Sample Survey Organisation, Government of India.
Department of Statistics. (1987). *Tables with notes on own-account trading enterprises and non-directory trading establishments covering some major aspects (employment, expenditure, valued added etc., NSS 34th round, July 1979–June 1980, Report number 346/1.* National Sample Survey Organisation, Government of India.
Department of Statistics. (1989). *Tables with notes on own-account trading enterprises and non-directory trading establishments—All-India, NSS 41st round, July 1985–June 1986, Report number 366/1.* National Sample Survey Organisation, Government of India.
Department of Statistics. (1991). *A report on living conditions of tribals and non-tribals of tribal areas, NSS 44th round, July 1988–June 1989, Report number 380.* National Sample Survey Organisation, Government of India.
Department of Statistics. (1996). *Level and pattern of consumer expenditure, NSS 50th round, July 1993–June 1994, Report number 402.* National Sample Survey Organisation, Government of India.
Department of Statistics. (1998). *Maternal and child health care in India, NSS 52nd Round, July 1995–June 1996, Report number 445.* National Sample Survey Organisation, Government of India.
Manna, G. C. (2024, April 9). A new methodology with some issues. *The Hindu*, Opinion.
Ministry of Finance. (1952). *The National Sample Survey, General report No. 1 on the first round, October 1950—March 1951.* The Department of Economic Affairs, Government of India.
MoSPI. (2001a). *Concepts and definitions used in NSS, Golden jubilee publication.* National Sample Survey Organisation, Ministry of Statistics and Program Implementation, Government of India.
MoSPI. (2001b). *Level and pattern of consumer expenditure in India 1999–2000, NSS 55th round, July 1999–June 2000, Report number 457.* National Sample Survey Organisation, Ministry of Statistics and Program Implementation, Government of India.
MoSPI. (2003). *Disabled persons in India, NSS 58th round, July–December 2002, Report number 485.* National Sample Survey Organisation, Ministry of Statistics and Program Implementation, Government of India.
MoSPI. (2004). *Evolution of the sample design in NSS from 1st to 55th round—A critical review, NSS technical note No. 2/03.* National Sample Survey Organisation, Ministry of Statistics and Program Implementation, Government of India.
MoSPI. (2005a). *Household consumer expenditure and employment-unemployment situation in India, NSS 59th round, January–December 2003, Report number 490.* National Sample Survey Organisation, Ministry of Statistics and Program Implementation, Government of India.
MoSPI. (2005b). *Household indebtedness in India as on 30.06.2002, NSS 59th round, January–December 2003, Report number 501.* National Sample Survey Organisation, Ministry of Statistics and Program Implementation, Government of India.
MoSPI. (2005c). *Income, expenditure and productive assets of farmer households, NSS 59th round, January–December 2003, Report number 497.* National Sample Survey Organisation, Ministry of Statistics and Program Implementation, Government of India.
MoSPI. (2006a). *Level and pattern of consumer expenditure, 2004–05, NSS 61st round, July 2004–June 2005, Report number 508.* National Sample Survey Organisation, Ministry of Statistics and Program Implementation, Government of India.
MoSPI. (2006b). *Livestock ownership across operational land holding classes in India, 2002–03, NSS 59th Round, January–December 2003, Report number 493.* National Sample Survey Organisation, Ministry of Statistics and Program Implementation, Government of India.
MoSPI. (2010a). *Domestic tourism in India, 2008–09, NSS 65th round, July 2008–June 2009, Report number 536.* National Sample Survey Organisation, Ministry of Statistics and Program Implementation, Government of India.

MoSPI. (2010b). *Housing condition and amenities in India 2008–09, NSS 65th round, July 2008–June 2009, Report number 535.* National Sample Survey Office, Ministry of Statistics and Program Implementation, Government of India.

MoSPI. (2010c). *Migration in India, 2007–2008, NSS 64th round, July 2007–June 2008, Report number 533.* National Sample Survey Office, Ministry of Statistics and Program Implementation, Government of India.

MoSPI. (2014a). *Employment and unemployment situation in India, NSS 68th round, July 2011–June 2012, Report number 554.* National Sample Survey Office, Ministry of Statistics and Program Implementation, Government of India.

MoSPI. (2014b). *Level and pattern of consumer expenditure 20ll-12, NSS 68th round, July 2011–June 2012, Report number 555.* National Sample Survey Office, Ministry of Statistics and Program Implementation, Government of India.

MoSPI. (2018). *Economic characteristics of unincorporated non-agricultural enterprises (excluding construction) in India, Report number 582, NSS 73rd round, July 2015–June 2016.* National Sample Survey Office, Ministry of Statistics and Program Implementation, Government of India.

MoSPI. (2019). *Technical report on services sector enterprises in India, NSS 74th round, July 2016–June 2017, Technical report No.TR (74/2.35).* National Sample Survey Office, Ministry of Statistics and Program Implementation, Government of India.

MoSPI. (2023). *Annual report, Periodic labour force survey (PLFS), July 2022–June 2023.* National Sample Survey Office, Ministry of Statistics and Program Implementation, Government of India.

MoSPI. (2024a). *Comprehensive annual modular survey, 2022–23, NSS 79th round, July 2022–June 2023, Report number 590.* National Sample Survey Office, Ministry of Statistics and Program Implementation, Government of India.

MoSPI. (2024b). *Operational and economic characteristics, Annual survey of unincorporated sector enterprises (ASUSE): 2021–22, April 2021–March 2022.* National Sample Survey Office, Ministry of Statistics and Program Implementation, Government of India.

MoSPI. (2024c). *Survey on household consumption expenditure: 2022–23 (HCES: 2022–23), August 2022–July 2023, Report number 591.* National Sample Survey Office, Ministry of Statistics and Program Implementation, Government of India.

MoSPI. (2025a). *Diamond jubilee publication, Household surveys of NSS, A journey of 75 years.* Ministry of Statistics and Program Implementation, Government of India.

MoSPI. (2025b). *Diamond jubilee publication, Enterprise surveys of NSS, A journey of 75 years.* Ministry of Statistics and Program Implementation, Government of India.

MoSPI. (2025c). *Periodic Labour Force Survey (PLFS), Changes in 2025. Ministry of Statistics and Programme Implementation*, Government of India.

Murthy, M. N. and Roy, A. S. (1975). Development of the sample design of the Indian National Sample Survey during its first twenty-five rounds. *Sankhya 37*(Series C, Pt. 1), 1–42.

The Cabinet Secretariat. (1958). *Report on household transport operations, National Sample Survey, Number 13, Seventh round, October 1953–March 1954.* Government of India.

The Cabinet Secretariat. (1959). *Report on employment and unemployment, National Sample Survey, Number 16, Ninth round: Preliminary, May–November 1955.* Government of India.

The Cabinet Secretariat. (1962). *Technical paper on sample design, National Sample Survey, Number 70, Fourteenth round, July 1958–June 1959.* Government of India.

The Cabinet Secretariat. (1966). *Tables with notes on land utilisation survey and crop-cutting experiments, National Sample Survey, Number 106, Seventeenth round, July 1961–June 1962.* Government of India.

The Cabinet Secretariat. (1968). *Technical paper on sample design, National Sample Survey, Number 125, Nineteenth round, July 1964–June 1965.* Government of India.

The Department of Economic Affairs. (1954). *Technical paper on some aspects of the development of the sample design by D.B. Lahiri, Indian Statistical Institute, Calcutta.* National Sample Survey, Number 5, March. Ministry of Finance, Government of India.

Chapter 4
Major Changes in Sample Design

After having described the sample design for some of the rounds since the inception of the NSS till now in Chapter 3, a summarization of the major changes in the sample design of the NSS over time may be useful for the readers. This chapter focuses on the same.

4.1 Number of Stages of Sampling and the Units at Various Stages

Leaving aside the crop surveys in some initial rounds (starting from the 10th round), the NSS has conducted either household surveys or enterprise surveys[1] following a stratified multi-stage design with households/enterprises as the ultimate sampling units (USUs). In its various rounds, the NSS has mostly treated the villages/urban blocks as the first-stage units (FSUs) and households/enterprises as the second-stage units (SSUs). However, for the large sample villages/urban blocks, there has generally been an intermediate stage of sampling where such villages/blocks are divided into a few segments, termed as hamlet-groups (HGs)/sub-blocks (SBs) and a random sample of them is selected for preparing the frame of households or enterprises as the case might be. Thus, in such situations, the design becomes a three-stage one with villages/blocks as the FSUs, HGs/SBs as the SSUs and households/enterprises as the third-stage units/USUs.

[1] Although the Indian NSS has used the terminology 'enterprise' in its various surveys, actually these surveys considered 'establishment' (i.e., the unit located in a single physical location) as the ultimate sampling unit. For keeping consistency with the NSS terminology, this book uses the terms enterprise (which will consist of many branches/establishments of the enterprise for multi-establishment enterprises) and establishment interchangeably. See MoSPI (2001) for concepts and definitions used in NSS. See also Murthy (1967) and Murthy and Roy (1975) for discussions on sampling theory and development of sample design.

G C Manna, *75 Years of the Indian National Sample Survey*, India Studies in Business and Economics, https://doi.org/10.1007/978-981-96-7320-9_4

It is important to mention that in only a very few initial rounds, instead of villages/urban blocks being treated as the FSUs, they were considered as the SSUs since the complete list of villages and urban blocks of all the towns as per the census was not available at the time of sampling. For example, in NSS 4th to 7th rounds, tehsils in rural areas were the FSUs. Similarly, towns in urban areas formed the FSUs in NSS 3rd to 8th rounds.

To avoid the intermediate stage of sampling of HGs/SBs in the large villages/blocks and/or reduce the extent of such segment formation/selection, the concept of formation of 'sub-units' (SUs) was introduced in NSS 76th round (July–December 2018; see MoSPI (2019c)). As per this procedure being adopted in most of the surveys since NSS 76th round except the PLFS, the villages with census 2011 population of 1,000 or more and large UFS blocks with 250 or more households are notionally divided into a certain number SUs with such SUs being treated as the FSUs. However, the smaller villages/UFS blocks with size lower than the above threshold continue to remain as the FSUs.

It is also important to mention that a uni-stage sampling (as against the multi-stage sampling) was adopted only in the 74th round of the NSS (see MoSPI, 2019b) which was a dedicated enterprise-based survey of bigger units where the list frame of such units was developed based on the sixth EC frame/Business Register comprising enterprises employing 10 or more workers and the frame of companies accessed from the Ministry of Corporate Affairs.

4.2 Sampling Frame of FSUs

In rural areas, the list of villages as per the latest population census has mostly served as the sampling frame of FSUs. In the enterprise surveys also, the same has been the case with the only change that in the NSS rounds conducted after the introduction of economic census (EC) in 1977, the villages have been mostly selected after duly stratifying them based on the number of enterprises/workers as per the EC.

As regards the urban areas, the latest list of census enumeration blocks (CEBs) served as the frame of FSUs in most of the NSS rounds till NSS 15th round (1959–60). In the subsequent NSS rounds, with the introduction of the scheme of UFS and its regular updating roughly once in five years, the latest available list of UFS blocks has generally been used as the sampling frame of FSUs. In the case of enterprises surveys, the list of urban blocks giving information on number of enterprises/workers by industry as per the EC has been mainly used as the sampling frame for selection of FSUs, i.e., urban blocks.

4.3 Use of List Frame Supplementing the Area Frame

In its various enterprise surveys also, the NSS has used the 'area frame', i.e., the list of villages and urban blocks with count of number of enterprises and workers by industry at the village/block-level based on the EC as the sampling frame of FSUs. However, there were some exceptions in 62nd (2005–06) and 63rd round (2006–07)[2] enterprises surveys on unorganized manufacture and services sector respectively where the area frame was supplemented with the list frame of bigger enterprises. To illustrate, in the 62nd round (MoSPI, 2007), the list frame (for urban areas only) comprised about 8000 big registered manufacturing units (not covered in the Annual Survey of Enterprises) identified based on the data of the census of manufacturing enterprises conducted by Development Commissioner of Small Scale Industries (SSI) in 2003. Such units were the ones whose gross value of output in 2001 were more than six times the average output of all urban SSI units and at the same time which employed 6 or more workers. Similarly, in the NSS 63rd round (MoSPI, 2009), the said list frame included 998 service sector companies across India. For developing this list fame, the respective lists of financial sector enterprises and other service sector enterprises were received from the RBI and the Ministry of Company Affairs. In both these rounds, all the units of the list frame were surveyed on a complete enumeration basis.

4.4 Stratification of FSUs

After the commencement of the NSS in the year 1950, in the initial few rounds, rural strata were formed by grouping contiguous districts within a State. However, since NSS 28th round (October 1973–June 1974), each district has been considered as a separate stratum in the rural areas except that PLFS (2017–18 and onwards), Time Use Survey[3] (TUS: 2019 & 2024) as well as ASUSE-treated NSS Region as a basic stratum and the HCES: 2022–23/2023–24 considered each State/UT as the basic stratum. The consideration of district as a stratum facilitates generation of the district-wise estimates from the NSS data. With the adoption of the NSS Region (or State/UT in the HCES) as the stratum in the surveys mentioned above, the district-wise estimates may be difficult to arrive at since some of the districts may not be represented in the sample, or even if they are represented, the sample size may not be adequate in some of the districts.

So far as urban area is concerned, for a considerable period of time, the urban strata comprised the towns according to their population size with strata having been formed either within the country or within each 'census zone' or within a State

[2] MoSPI (2007, 2009).

[3] See MoSPI (2020) for TUS:2019; for TUS:2024, see https://pib.gov.in/PressReleasePage.aspx?PRID=2106113. For PLFS: 2017–18, see MoSPI (2019a). For HCES 2022–23, see MoSPI (2024a). For ASUSE 2021–22, see MoSPI (2024b).

(see Chapter 2 for details). Subsequently, during NSS 42nd round (1986–87) till 60th round (January–June 2004), urban strata were formed by grouping towns into some size classes according to their population within each NSS region. Keeping in view the demand for district level estimates, from NSS 61st round (July 2004–June 2005), NSS has considered each district (urban part) as a separate urban stratum (like the practice adopted for rural areas) barring the following exceptions: (i) NSS 65th round survey on domestic tourism, housing condition and slum (2008–09), PLFS, TUS (2019 & 2024), as well as ASUSE treating each NSS region as a basic stratum with further stratification of the towns within each NSS region according to their population and (ii) HCES: 2022–23/2023–24 considering each State/UT cross-classified by population size classes of towns as the basic stratum. Box 4.1 summarizes the changes in the formation of basic strata for sampling of FSUs in different rounds/surveys during the last two decades.

4.5 First-Stage Strata of Affluent Areas

There has been a general feeling that the surveys on household consumer expenditure (HCE) in the NSS have the deficiency in capturing adequate number of rich/affluent households through its sample. To guard against the same, the NSS has attempted the representation of relatively affluent households in the sample at the time of sampling of USUs, i.e., households by forming a separate stratum of such households at the listing stage. Such a practice has been followed in all the surveys on HCE since NSS 43rd round (July 1987–June 1988) till now. However, for the first time in the NSS 43rd round, steps were also taken to see that adequate number of FSUs (i.e., UFS blocks) located in the affluent areas are in the sample. This was done by further stratifying urban strata of towns with population of 400,000 or more into two groups/strata: (i) UFS blocks of the towns located in affluent area and (ii) the remaining UFS blocks of the stratum, with a higher allocation/weight given to the strata of affluent area blocks.[4] The feedback on list of such UFS blocks located in affluent areas was provided by the field offices of FOD, NSSO. Such a procedure of formation of strata of affluent area blocks was also followed in NSS 50th round (1993–94) and HCES: 2022–23/2023–24. It may be mentioned that the list of 'area type' codes prevalent in the UFS earlier (like UFS blocks located in bazar area, industrial area, etc.) has subsequently been modified by including in the list a dedicated code for the blocks located in the affluent areas with a view to using this information for stratification of FSUs, if required.

[4] Urban strata in NSS 43rd round (see Department of Statistics (1990)) were formed within each NSS region as follows: all towns with population less than 50,000 (stratum 1); all towns with population of 50,000 to less than 200,000 (stratum 2); all towns with population of 200,000 to less than 400,000 (stratum 3); affluent area blocks of towns with population of 400,000 to less than 1 million (stratum 4); remaining blocks of towns with population of 400,000 to less than 1 million (stratum 5); affluent area blocks of a city with population of 1 million or more (stratum 6); remaining blocks of a city with population of 1 million or more (stratum 7).

Box 4.1 Changes in the formation of basic strata during the last two decades

NSS Round Number/Survey	Basic Strata[5] (for sampling of FSUs)	
	Rural	Urban
61*, 62, 63, 64	District	District; Each M + City a separate stratum
65	District	NSS Region X Town size class; Each M + City a separate stratum
66, 67, 68, 69, 70, 71	District	District; Each M + City a separate stratum
72	District	District; Each city having a population of 4 lakh or more a separate stratum
73	District	District; Each 1.5M + City a separate stratum
74	(The round, being devoted to a survey of bigger enterprises, used a list frame of enterprises for a uni-stage sampling)	
75, 76, 77, 78, 79, 80	District	District; Each M + City a separate stratum
PLFS: 2017–18 and Onwards; TUS: 2019 and 2024	NSS Region [PLFS 2025 considers district as the stratum in most part of the country and the NSS region as stratum in the remaining part]	NSS Region X Town size class; Each 1.5M + City a separate stratum [PLFS 2025 adopts the same stratification as in rural with each M+ City being treated as a separate stratum]
ASUSE: 2021–22 and Onwards	NSS Region	NSS Region X Town size class; Each M + City a separate stratum
HCES: 2022–23 and 2023–24	State/UT	State/UT X Town size class; Each M + City a separate stratum

Note *NSS 61st round was conducted during July 2004–June 2005
(M + City means a city with a population of 1 million or more as per the census; 1.5 M + City meaning a city with a population of 1.5 million or more as per the census)
Source Various NSS rounds, PLFS, TUS, ASUSE and HCES reports

4.6 Method of Sampling of FSUs

There have been changes from time to time in the method of sampling of FSUs in the NSS. These changes in the sampling procedures have been summarized in Box 2.2 of Chapter 2. So far as sampling of FSUs in the rural areas is concerned, in most of

[5] The basic strata as presented in Box 4.1 implies that it is theoretically possible to arrive at the estimates for each basic stratum. However, the level of precision of the estimates at the stratum level is a separate issue.

the rounds/surveys they have been selected by PPSWR followed by the method of circular systematic sampling with PPS[6] as the second most frequent method. It is also important to mention in this context that during certain early rounds of the NSS, the method of cluster sampling of villages was adopted in respect of the North-Eastern Region, particularly for the State of Arunachal Pradesh, mainly with a view to minimize the travel. To illustrate, in the NSS 32nd Round (July 1977–June 1978) devoted to the second quinquennial survey on household consumer expenditure and employment—unemployment, for the North-Eastern Region, a sample of 'nucleus' villages was selected first systematically with equal probability from each basic stratum (i.e., district or part of a district in case of larger districts in terms of population). Thereafter, clusters of villages were formed around the selected nucleus villages for the purpose of survey. A cluster consisted of 4 villages in general but the number did not exceed 6. In the State of Manipur, however, such a cluster sampling was not resorted to and instead the required number of sample villages was selected systematically with equal probability. In some of the subsequent rounds (e.g., NSS round numbers 48, 50 and 52), this method of cluster sampling of villages was confined only to the State of Arunachal Pradesh.

As regards the sampling of FSUs in the urban areas, no specific method of sampling had an edge in terms of its frequency of adoption in the NSS. For example, the PLFS series has been adopting PPSWR technique as against the method of SRSWOR used in the other recent surveys since NSS 77th round. It is of interest to note that the method of circular systematic sampling (with equal probability or with PPS) retaining the geographical arrangement of the villages/towns as per census was adopted continuously in all the NSS rounds from 14th round till 27th round and also in quite a few rounds later. The purpose was to ensure geographical spread of the sample FSUs within each stratum.

4.7 Selection of Segments from Large FSUs

In most of the NSS rounds, the sample villages have been selected by PPS with village population as per the census as the size measure. Such a procedure generally leads to selection of large villages more in number. Since the preparation of frame of USUs in such large villages is a time-consuming exercise, to control the workload at the listing stage, the common practice in the NSS has been to select a sample of segments of the village/block (either one or two) for preparing the frame of USUs.

There have been some changes in the method of selection of such segments in the NSS. For the first time, the method of segment selection in large villages was introduced in NSS 9th round (May–November 1955) when one natural hamlet of the

[6] The method of circular systematic sampling with PPS adopted in many NSS rounds retains the advantage of PPS sampling with the large FSUs (in terms of population, etc.) having higher chances of inclusion in the sample. At the same time, this method ensures geographical spread of the sample FSUs within a stratum since the FSUs are drawn with a 'sampling interval' after retaining the arrangement of the FSUs in the frame as per their ordering according to geographical locations.

village was selected by PPS with approximate hamlet population as the size measure. Such a procedure continued till NSS 13th round. The procedure was modified in 14th round (July 1958–June 1959) when instead of selecting one natural hamlet, the large sample villages were divided into a few HGs of roughly equal size and one HG was selected at random. A similar method of segmentation was adopted for large sample blocks also which were divided into a few SBs with one SB selected at random for the purpose of listing. This procedure was followed in almost all the rounds up to 44th round (1988–89). Subsequently, the procedure of selecting 2 HGs/SBs has been adopted in most of the NSS rounds.

An important and significant change in the method of sampling of HG/SBs was incorporated for the first time in NSS 41st round (1985–86) survey on trading enterprises.[7] In this round, before dividing the large villages/blocks (with population of 1,200 or more) into a specified number of HGs/SBs, an 'area of concentration' (like market area or bus stop) with the maximum concentration of trading enterprises within the whole of FSU was carved out provided such a compact area contained at least 10 trading enterprises. The remaining area of the FSU was as usual divided into a few HGs/SBs out of which one was selected at random. The listing exercise and sampling of enterprises was carried out separately in the area of concentration as well as in the randomly selected HG/SB. A more or less similar method is being adopted in most of the recent surveys with the only change that instead of carving out such an area of concentration within the large FSU, the HG/SB having maximum population (or enterprises), depending upon whether it is a survey of households or enterprises, is selected with certainty and one more HG/SB is selected at random from among the remaining ones.

4.8 Sampling of USUs

For a considerable period of time till NSS 56th round (2000–01), use of systematic sampling (linear or circular) was the most common method in the NSS for sampling of USUs, i.e., households/enterprises. Subsequently, all the NSS rounds/surveys from NSS 57th round (July 2001–June 2002) have adopted SRSWOR method of sampling to select the sample USUs.

To illustrate, sample households in NSS 14th round were selected systematically with a random start. In NSS 19th round, sample households were selected by linear systematic sampling after arranging the households based on household size and means of livelihood. In the 27th round survey on employment–unemployment (E-U) and HCE, listed households in the FSU were stratified into 2 sub-strata: self-employed households (sub-stratum 1) and households which were not self-employed (sub-stratum 2). Thereafter, households of sub-stratum 1 were arranged into 2 classes: self-employed in manufacturing and self-employed in other activities.

[7] See Department of Statistics (1989).

Finally, required number of sample households were selected from each sub-stratum by linear systematic sampling.

Subsequently, there was a shifting from linear systematic sampling to the method of circular systematic sampling. In NSS 43rd round survey on E-U and HCE, listed households in the sample FSUs were stratified into 2 strata: those comprising relatively affluent households (stratum 1) and the rest (stratum 2). Thereafter, households of stratum 2 in the rural FSUs were grouped into 3 classes: self-employed in non-agriculture (class 1), rural labour (class 2) and other households (class 3) with further arrangement of households in class 3 according to 5 land possessed classes. In the urban FSUs, listed households of stratum 2 (i.e., with MPCE below certain threshold) were grouped into 4 means of livelihood classes, i.e., self-employed, regular wage/salaried, casual labour and others. Finally, a circular systematic sample of households was selected from each stratum of both rural and urban FSUs. Exactly similar method was also followed in the subsequent two quinquennial surveys on E-U and HCE, i.e., during NSS 50th and 55th rounds.

In the survey of manufacturing enterprises of NSS 51st and 56th rounds, manufacturing enterprises in the FSU were stratified into 3 strata of OAMEs, NDMEs and DMEs[8] and the required number of sample enterprises from each stratum were selected circular systematically after arranging them according to broad industry codes.

The method of systematic sampling was in place primarily with a view to having in the ultimate sample the proportionate representation of USUs of different categories as per the variable used for the arrangement of the USUs.

4.9 Self-Weighting Design

As mentioned in Chapter 2, for quite a few NSS rounds since fourth round, NSS design was made 'self-weighting' where the 'multiplier' attached to all the sample households for a given domain of interest (e.g., State) was kept a constant number for derivation of estimates easily. However, because of certain disadvantages of the design including its necessity for selection of unequal number of USUs per FSU, the method was subsequently given up.

[8] OAMEs: Own-Account Manufacturing Enterprises (i.e., without any hired workers); NDMEs: Non-Directory Manufacturing Establishments (i.e., with at least one hired worker and total number of workers being 5 or less); DMEs: Directory Manufacturing Establishments (i.e., with at least one hired worker and total number of workers being 6 or more). See https://mospi.gov.in/sites/default/files/publication_reports/design_golden.pdf.

4.10 Use of Rotational Panel Design

In the NSS, sample FSUs (villages/urban blocks) in each round are drawn independently even for the repetitive surveys on the same subject of enquiry. However, the urban blocks in the PLFS series from 2017–18 till December 2024 have been drawn following the rotational panel design with details discussed in Chapter 3 where sample blocks for each quarter are selected in the form of 4 panels of equal number of sample blocks in each such that one of these panels is introduced in the quarter for the first time with other 3 panels being common with the previous quarter. This rotational panel design was introduced in the urban component of the PLFS (2017–18 till December 2024) with a view to having better estimates of change between two quarters. PLFS 2025 has extended the rotational sampling design to rural areas also and it has adopted a monthly rotational scheme (MoSPI, 2025).

References

Department of Statistics. (1989). *Tables with notes on own-account trading enterprises and non-directory trading establishments—All-India, NSS 41st round, July 1985–June 1986, Report number 366/1.* National Sample Survey Organisation, Government of India.

Department of Statistics. (1990). *Results of the fourth quinquennial survey on employment and unemployment (All India), Sarvekshana, NSS 43rd round, July 1987–June 1988.* National Sample Survey Organisation, Government of India.

MoSPI. (2001). *Concepts and definitions used in NSS, Golden jubilee publication.* National Sample Survey Organisation, Ministry of Statistics and Programme Implementation, Government of India.

MoSPI. (2007). *Operational characteristics of unorganised manufacturing enterprises in India, NSS 62nd round, July 2005–June 2006, Report number 524.* National Sample Survey Organisation, Ministry of Statistics and Programme Implementation, Government of India.

MoSPI. (2009). *Service sector in India (2006–07): Economic characteristics of enterprises, NSS 63rd round, July 2006–June 2007, Report number 529.* National Sample Survey Organisation, Ministry of Statistics and Programme Implementation, Government of India.

MoSPI. (2019a). *Annual report, Periodic Labour Force Survey (PLFS), July 2017–June 2018.* National Statistical Office, Ministry of Statistics and Programme Implementation, Government of India.

MoSPI. (2019b). *Technical report on services sector enterprises in India, NSS 74th round, July 2016–June 2017, Technical report no.TR (74/2.35).* National Sample Survey Office, Ministry of Statistics and Programme Implementation, Government of India.

MoSPI. (2019c). *Persons with disabilities in India, NSS 76th round, July –December 2018, Report number 583.* National Statistical Office, Ministry of Statistics and Programme Implementation, Government of India.

MoSPI. (2020). *Time use in India–2019, Time use survey, January–December 2019.* National Statistical Office, Ministry of Statistics and Programme Implementation, Government of India.

MoSPI. (2024a). *Survey on household consumption expenditure: 2022–23 (HCES: 2022–23), August 2022–July 2023, Report number 591.* National Sample Survey Office, Ministry of Statistics and Programme Implementation, Government of India.

MoSPI. (2024b). *Operational and Economic Characteristics, Annual Survey of Unincorporated Sector Enterprises (ASUSE) 2021–22*, April 2021–March 2022. National Sample Survey Office, Ministry of Statistics and Programme Implementation, Government of India.

MoSPI. (2025). *Periodic Labour Force Survey (PLFS), Changes in 2025.* National Statistics Office, Ministry of Statistics and Programme Implementation, Government of India.
Murthy, M. N. (1967). *Sampling theory and methods.* Statistical Publishing Society.
Murthy, M. N. and Roy, A. S. (1975). Development of the sample design of the Indian National Sample Survey during its first twenty-five rounds. *Sankhya 37* (Series C, Pt. 1), 1–42.

Chapter 5
Pilot Surveys

The NSSO conducted two important pilot surveys with a moderate sample size. The first one was on income, consumption and savings conducted during September 1983–December 1984, while the second one (January–June 2000) was aimed at finding out appropriate reference periods for measuring household consumption. In between, another pilot survey covering five States was conducted in 1986-87 to test the appropriateness of 'last month' as the reference period used in most of the surveys on household consumption. This pilot survey used last month and last week as the two alternate reference periods for data collection to compare the estimates as per these alternative reference periods with the one based on 'data collected through daily visits to sample households for three or four consecutive days'. However, results/findings based on this survey are not available in public domain. Another pilot survey conducted by the NSSO on a limited scale during the nineties, with results not available in public domain, was to experiment with an alternative approach of data collection for estimation of the gross value added (GVA) by the enterprises.

5.1 Pilot Survey of Income, Consumption and Savings (September 1983–December 1984)

Studies on household income and income-based inequality in India have been hindered to a great extent due to a lack of data on income. The only direct enquiries on household income at the all-India level were conducted by the National Council of Applied Economic Research (NCAER). However, the estimates from these surveys are not free from limitations.

On an experimental basis, the NSS made some attempts in its 9th to 14th rounds to collect data on household income along with data on other subjects of enquiry including the household consumer expenditure. Subsequently, NSS collected data on income, consumer expenditure and other subjects through an integrated schedule

G C Manna, *75 Years of the Indian National Sample Survey*, India Studies in Business and Economics, https://doi.org/10.1007/978-981-96-7320-9_5

in the 19th and 24th rounds also. However, estimates were affected due to a lengthy schedule of enquiry. Recently, during NSS 59th, 70th and 77th rounds, data on household income was collected but the target group in these surveys were only the farmer/agricultural households residing in rural areas.

With a view to evolving a proper methodology to collect data on income through the NSS, a pilot survey of income, consumption and savings was conducted by the NSSO during 1983-84 covering both rural and urban areas of 5 States namely Haryana, Maharashtra, Odisha, Tamil Nadu and Uttar Pradesh along with the metropolitan cities of Calcutta and Delhi. Based on this pilot survey, NSSO released two reports, i.e., NSS Report Number 390 (being Part I of the report on pilot survey) in August 1993 and thereafter NSS Report Number 398 (Part II of the report) in September 1995.

5.1.1 Sampling Frame and Stratification of FSUs

The pilot survey adopted a stratified two-stage design with 1981 census villages and the latest available UFS blocks comprising the FSUs in rural and urban areas respectively and households as the SSUs. However, as usual, an intermediate stage of sampling was the selection of an HG/SB for the large FSUs with a population of 1,200 or more.

In the rural areas, each NSS region or a group of NSS regions formed a stratum while in the urban areas each metropolitan city of Bombay, Calcutta, Delhi and Madras formed a separate stratum and the remaining cities and towns within a State were grouped into 2 strata: towns with population of 200,000 and above (stratum 1) and remaining towns (stratum 2).

5.1.2 Sample Size of FSUs, Their Allocation and Sampling Procedure

Altogether 100 villages and 80 UFS blocks were selected for the survey. This overall sample size of FSUs for each rural and urban was first allocated to the States in proportion to their respective shares in the total rural/urban population. Thereafter, rural/urban sample size of a State was further allocated to the rural/urban strata of the State considering their relative shares in population with stratum level allocations adjusted to multiples of 4 to facilitate the selection of sample FSUs from each stratum in the form of 2 independent sub-samples and 2 sub-rounds. The State/City-wise sample allocations (see Box 5.1) were as under[1]:

[1] Department of Statistics (1993).

Box 5.1 Allocation of sample villages and blocks to different states in the pilot survey

State	Rural	Urban		
		Other than metropolitan cities	Metropolitan city	Corresponding sample allocation
Haryana	12	8	Calcutta	4
Maharashtra	20	8	Bombay	8
Odisha	16	8	Delhi	4
Tamil Nadu	20	8	Madras	8
Uttar Pradesh	32	24		
All	**100**	**56**		**24**

Source Department of Statistics (1993)

From each stratum, the required number of sample FSUs was selected circularly systematically with probability proportional to population in case of rural and with equal probability in case of urban, in the form of two independent sub-samples.

5.1.3 *Stratification and Sampling of Households*

Listed households of each FSU/HG/SB were grouped into three sub-strata by jointly considering whether there was at least one member in the household having a white collar job and the quantum of land possessed by the household in case of rural and by considering the household MPCE in case of urban. For this purpose, in rural areas, households other than those having members with white collar job were first arranged in the ascending order of land possessed, followed by households having members with white collar job. The first 30% of the households, the next 55% and remaining 15% of the households respectively formed sub-strata 1, 2 and 3. In the urban areas, households were arranged in ascending of the MPCE. Thereafter, the first 30%, the next 60% and the last 10% of the households in the arranged frame respectively formed sub-strata 1, 2 and 3. Finally, from both rural and urban FSUs/HGs/SBs, 9 households each from each of sub-strata 1, 2 and 6 households from sub-stratum 3 were selected by circular systematic sampling.

5.1.4 *Canvassing the Household Schedules of Enquiry*

There were two detailed schedules in the pilot survey namely Schedule 1.1A: Income; and Schedule 1.1B: Consumption and Savings. The selected households were divided into 3 sets (see Box 5.2) as per their order of selection as follows:

Box 5.2 Order of selection of households in the pilot survey

Sub-stratum Number	Order of selection numbers of the sample households allotted to different sets		
	Set I	Set II	Set III
1	1, 4, 7	2, 5, 8	3, 6, 9
2	1, 4, 7	2, 5, 8	3, 6, 9
3	1, 4	2, 5	3, 6

Source Department of Statistics (1993)

In the sample households belonging to Set I, only income data were collected by canvassing Schedule 1.1A. In the households of Set II, data on consumption expenditure and savings were collected through Schedule 1.1B while in Set III households both the schedules were canvassed to collect information on income, consumption expenditure as well as savings. The rationale for doing so was to judge the income estimate, which is generally weak, against the estimate of consumption expenditure plus savings. The latter was believed to be closely approximating the true income.

Data were collected by performing two visits to each household. In the States of Maharashtra, Tamil Nadu and Uttar Pradesh, data were collected during September 1983–August 1984 while in the other regions, household interviews took place during January–December 1984. For income and savings data, the reference period was six months of either January to June or July to December of a year. By adding the data for two reference periods, the yearly figure of income and the savings of a hold were derived. That way, the one-year reference period was January–December 1983 for the States of Maharashtra, Tamil Nadu and Uttar Pradesh and July 1983–June 1984 for the other States/regions. For the household consumer expenditure, a reference period of 30 days preceding the date of survey was adopted for collecting the relevant data during each visit.

5.1.5 Major Findings and Recommendations

As documented in NSS Report Number 398, for the rural areas of five States together, the average value of income was found to be lower by 30% as compared to the average value of consumption plus savings for Set III households (Department of Statistics, 1995). Even the larger difference was noticed between the average income (Rs. 4,825) for Set I households and the average value of consumption plus savings (Rs. 8,075) for Set II households. On the contrary, for the urban areas including the metropolitan cities, the said difference between the two was not at all significant. Some of the major limitations of the pilot survey including the recommendation as per the said report were as follows (Department of Statistics, 1995):

i. Reported data on income and savings was not of high quality;

ii. Sample size was found to be inadequate for highly variable items like savings and its components;
iii. The schedules of enquiry did not 'seem to have been well designed' to collect data on non-monetized components of income and consumption; and
iv. 'One needs to conduct further full-scale pilot surveys for finally arriving at a satisfactory methodology for household income surveys'.

5.2 Pilot Survey on Suitability of Different Reference Periods for Measuring Household Consumption (January–June 2000)

The data on household consumer expenditure have been covered in many rounds of the NSS since its inception. A due attention has been paid to the choice of reference period of the items of consumption in these surveys. In the quinquennial surveys on the subject during NSS 27th, 32nd, 38th, 43rd and 50th rounds, the last 30 days preceding the date of survey have been chosen as the reference period for all the items of household consumption. However, to eliminate the effect of seasonality, 'last 365 days' (in addition to last month) was added as an additional reference period for certain infrequently purchased item groups. These were clothing, footwear and durable goods in rounds 32, 38 and 43 with the addition of item groups of education and medical expenses (institutional) in NSS 50th round.

Subsequently, for comparative purposes of estimates based on the alternative reference periods, two types of schedules with varying reference periods were canvassed in two different sets of households in each FSU during NSS 51st to 54th rounds. While Schedule Type 1 had a uniform reference period of the last 30 days for all the items, Schedule Type 2 had a reference period of 'last 7 days' for food, pan, tobacco and intoxicants; 'last 30 days' for fuel and light, miscellaneous goods and services and medical (non-institutional); and 'last 365 days' for the item groups of clothing, footwear, durable goods, education and medical (institutional). As per NSS Report No. 447 (Department of Statistics, 2000), the week-based estimates of food, pan, tobacco and intoxicants based on these surveys were about 30% higher than the month-based estimates. Under the above backdrop, the said pilot survey was carried out to take a proper call on the most appropriate reference period for the items of food, pan, tobacco and intoxicants.

5.2.1 Geographical Coverage and Sample Size of FSUs

The pilot survey covered both rural and urban areas of 5 States namely Haryana, Maharashtra, Orissa, Tamil Nadu and Uttar Pradesh. The survey was extended to the two big cities of Delhi and Kolkata in addition to Chennai and Mumbai belonging to the States of Tamil Nadu and Maharashtra. Due to resource constraints, only two NSS regions of Uttar Pradesh and one NSS region from each of the 4 remaining States were covered in the pilot survey. The regions were selected purposively keeping in view the region-wise poverty estimates by Dubey and Gangopadhyay (1998). The selected regions were Haryana: Western; Maharashtra: Inland Eastern; Orissa: Southern; Tamil Nadu: Inland; and Uttar Pradesh: Central & Eastern. The total sample size of FSUs considering all the States was 96 villages and 120 UFS blocks (48 blocks from 4 big cities and 72 blocks selected from the other urban areas).

5.2.2 *Broad Sample Design and Method of Sampling of FSUs*

A stratified two-stage design was adopted with 1991 census villages and the latest available UFS blocks in the urban sector as the FSUs and households as the USUs. An intermediate stage of sampling in large FSUs was selection of one HG/SB at random from among the HGs/SBs formed with roughly equal population content. In rural, the entire rural area of each NSS region was treated as a separate stratum. In case of urban, each of the four big cities constituted a separate stratum while the remaining urban areas of each region formed another stratum. The sample FSUs from each stratum were selected by circular systematic sampling with probability proportional to population in the rural areas and with equal probability in the urban areas—in the form of two independent sub-samples. For the purpose of sampling, FSUs in the frame of any rural stratum were arranged as per their arrangement in the 1991 census (MoSPI, 2002).

5.2.3 *Sampling of Households and the Schedules of Canvassing*

All the listed households in an FSU/HG/SB were arranged according to their 'means of livelihood' x land possessed in case of rural and according to MPCE in case of urban. Thereafter, a systematic sample of 16 households was selected for detailed enquiry. Selected 16 households were divided into 4 groups each comprising 4 consecutive households in the arrangement. Each group of 4 consecutive households was considered for canvassing four different schedules—one each for Schedule 1.0(M), Schedule 1.0(W), Schedule 1.0(D3) and Schedule 1.0(D4). Schedule 1.0(M) was canvassed in the first household of each group. In the three following households of the group, Schedule 1.0(W), Schedule 1.0(D3) and Schedule 1.0(D4) respectively were canvassed (MoSPI, 2002).

5.2.4 Details of Data Collection and the Rationale

Schedule 1.0(M) had the reference period of 'last 30 days' for food, pan, tobacco and intoxicants, fuel and light, miscellaneous goods and services including medical (non-institutional) but 'last 365 days' for five remaining item groups, i.e., clothing, footwear, durables, education and medical (institutional). Schedule 1.0(W) had 'last 7 days' as the reference period for food, pan, tobacco and intoxicants while for the rest of the items, the reference periods were the same as in Schedule 1.0(M). Schedule 1.0(D3) and Schedule 1.0(D4) were designed with 'yesterday' as the reference period for items of food, pan, tobacco and intoxicants only (MoSPI, 2002) with no provision of data collection for other items. Schedule 1.0(D3) and Schedule 1.0(D4) were respectively canvassed by visiting the selected households on three consecutive days and on four consecutive days.

As regards the data collection with last day as the reference period, to reproduce from NSS Report Number 475, for households selected for Schedule 1.0 (D3/D4), 'special efforts were made to collect accurate data through daily interviews and by measurement for key items, as far as possible. First, each household was supplied with notebook and dot pen and requested to keep a record of consumption of each spending member of the household. More importantly, each household was supplied with two transparent graduated containers with volume markers, one measuring 500 ml and the other 100 ml, and they were urged to measure out cereals and cereal products, pulses and pulse products and liquid milk consumed each day using these containers instead of their own containers' (MoSPI, 2002).

It was believed that with the above measures in place, the estimate of household MPCE based on the last day as the reference period would be near to the true value. Accordingly, between the two alternative reference periods of 'last 30 days' and 'last 7 days', the one giving MPCE estimate closer to the estimate based on the last day as the reference period may be considered as the appropriate reference period for the items of food, pan, tobacco and intoxicants.

5.2.5 Major Findings and Recommendations

NSS Report Number 475[2] describes in detail the main findings of the pilot survey including (a) item-wise estimates of quantity and value of consumption per person per 30 days by schedule type for some major food items (refer to Table 4 of the report); and (b) item-wise estimates of value of consumption per person per 30 days by schedule type for major food and non-food items (vide Table 5 of the report)—with the said estimates cross-classified by sector (i.e., rural and urban: metro/non-metro/combined) for each sub-sample. Some of the major findings and recommendations are reproduced as they are below from the report:

[2] MoSPI (2002).

i. The week-based estimates of expenditure or value of consumption are, by and large, appreciably higher than the corresponding month-based estimates and are generally much closer to the day-based estimates. This shows that 'last week' would be more appropriate than 'last month' in many cases.[3]
ii. It seems reasonable to conclude that, on the whole, for both rural and urban sectors, the 'last week' reference period would be more appropriate than the 'last month' reference period for many groups of items, e.g., vegetables, fruits (fresh) and spices, because the week-based estimates of the value of consumption appear to be closer to the corresponding day-based estimates, which are taken as standard. Similar conclusions may also be drawn for the 'all food' group, as a whole and for pan, tobacco and intoxicants. However, for some item groups including cereals and products, which are important for nutrition-oriented studies, 'last month' estimates look better or reasonable and so 'last month' could be retained as reference period for such groups.[4]
iii. A more serious consideration is that if one adopts last week as the reference period, one has to stagger the interviews evenly over the 52 weeks of the survey year and not merely over the 12 months of the survey year.[5]
iv. The finding that day-based estimates are appreciably higher than even the week-based estimates cannot be ignored. However, the adoption of 'day or yesterday' as a reference period for the nationwide surveys conducted by NSSO seems to be infeasible. The cost would be manifold, and cooperation of the kind received in this Pilot Survey from the informants cannot be guaranteed in repetitive surveys. Repeated interviews for the same household for three or four consecutive days and collection of data on the volume of consumption through containers would not be possible on any large-scale year after year.[6]

5.3 Study on Alternative Approach of Data Collection for Measuring GVA (Mid-Nineties)

Estimates of GVA per worker by industry for the unorganized enterprises based on the NSS are generally used to estimate the contribution of the unorganized/household sector in the overall GVA/gross domestic product (GDP). Traditionally, in its enterprise surveys, the NSS collects information on both receipts and operational expenses of the enterprises—which are used in the estimation of GVA—from the same sample enterprises. There has been a feeling that the NSS estimates of GVA per worker are underestimates. Accordingly, during the mid-nineties, as per the advice of the NSSO Governing Council, a pilot study was conducted by trying out an alternative approach

[3] Ibid.
[4] Ibid.
[5] Ibid.
[6] Ibid.

Box 5.3 Percentage difference of GVA between the production approach and factor income approach, NSS 67th round, July 2010–June 2011

Broad Activity	Percentage difference of GVA between the two approaches expressed as percentage of GVA based on the production approach								
	Rural India			Urban India			India		
	OAE	EST	ALL	OAE	EST	ALL	OAE	EST	ALL
(1)	(2)	(3)	(4)	(5)	(6)	(7)	(8)	(9)	(10)
MFG	2.32	1.16	1.75	1.15	0.52	0.69	1.75	0.69	1.08
TR	2.44	6.89	3.49	2.60	1.71	2.10	2.54	2.46	2.50
OS	1.82	0.54	1.29	1.66	0.90	1.16	1.73	0.82	1.20
ALL	2.21	2.20	2.21	2.02	1.09	1.43	2.10	1.32	1.67

Note MFG: Manufacturing TR: Trade OS: Other Services; OAE: Own Account Enterprises, EST: 'Establishment', i.e., units having hired workers
Source MoSPI (2013)

in which the information on receipts and operating expenses was collected respectively from two different sets of trading enterprises in the selected sample FSUs. Thereafter, the estimate of aggregate GVA was derived as the difference between estimated receipts and estimated operating expenses as obtained from the respective data of two different sets of enterprises. However, the estimate of GVA based on this new approach of data collection did not yield encouraging results. Accordingly, this procedure was not implemented in any enterprise survey and the traditional approach to collect data on receipts and operating expenses from the same set of enterprises as followed earlier continued in the subsequent NSS rounds as well. However, in some of the NSS rounds later on (e.g., NSS 67th and 73rd rounds), the direct question on 'net surplus' was also added to the schedule of enquiry facilitating estimation of the GVA as per the 'factor income approach'[7] in addition to the 'production approach' adopted in the surveys where GVA is calculated as the difference between the receipts and operating expenses.

It is of interest to mention that the factor income approach of GVA did not show any improvement. In fact, the alternative estimate of GVA as per the production approach yielded a somewhat higher estimate GVA (see Box 5.3) as compared to the factor income approach for most of the activities in NSS 67th round.

[7] In this approach, GVA is derived by adding the values corresponding to all the four factors of production namely emoluments to employees, rent payable, interest payable and net surplus. The first three items of factors of production normally form part of the schedule of enquiry in the enterprise surveys.

References

Department of Statistics. (1993). *Report on the pilot study of income, consumption and savings—Part I: Income, Methodological study, September 1983–December 1984, Report number 390.* National Sample Survey Organisation, Government of India.

Department of Statistics. (1995). *Report on the pilot study of income, consumption and savings—Part II, Methodological study, September 1983–December 1984, Report number 398.* National Sample Survey Organisation, Government of India.

Department of Statistics. (2000). *Choice of reference period for consumption data based on NSS 51st round, NSS 52nd round, NSS 53rd round, NSS 54th round, Report number 447.* National Sample Survey Organisation, Government of India.

Dubey, A., & Gangopadhyay, S. (1998). *Counting the poor: Where are the poor in India?* Department of Statistics, Government of India.

MoSPI. (2002). *Results of a pilot survey on suitability of different reference periods for measuring household consumption, January–June, 2000, Report number 475.* National Sample Survey Organisation, Ministry of Statistics and Program Implementation, Government of India.

MoSPI. (2013). *Economic characteristics of unincorporated non-agricultural enterprises (excluding construction) in India, NSS 67th round, July 2010–June 2011, Report number 549.* National Sample Survey Office, Ministry of Statistics and Program Implementation, Government of India.

Chapter 6
Some Global Survey Practices

Against the backdrop of the survey designs adopted in the NSS, it will be useful to see the survey practices of some other countries. This chapter focuses on the same in respect of two important household surveys namely the household expenditure survey and the labour force survey. However, before the survey practices on these two surveys by the other countries are discussed, Sect. 6.1 of this chapter first summarizes the survey methodology adopted in the NSS in respect of these two surveys so that the readers find it useful to have a comparative picture.

6.1 Methodology Adopted in India: Surveys on Household Consumer Expenditure and Employment–Unemployment[1]

Both the surveys on household consumer expenditure (HCE) and employment–unemployment (E-U) have a long history in the NSS. In fact, the subject of HCE was taken up in the very first round of the NSS (October 1950–March 1951) and the subject got covered in many succeeding NSS rounds. The survey on E-U was conducted for the first time in NSS 9th round (May–November 1955).[2] As a part of the 'quinquennial surveys', the subjects of both HCE and E-U have been covered in nine NSS rounds namely NSS 27th round (October 1972–September 1973), 32nd round, 38th round, 43rd round, 50th round, 55th round, 61st round, 66th round and 68th round (July 2011–June 2012). Apart from the quinquennial surveys on HCE and E-U, there were annual surveys on HCE on a 'thin sample basis' during NSS

[1] This section contains information from the MoSPI (2014a, 2014b). See also MoSPI (2013). Other sources include MoSPI (2024a, 2024b).

[2] See Statement 2.1 of Chapter 2 for the various surveys/NSS rounds in which the subjects of HCE and E-U were covered.

G C Manna, *75 Years of the Indian National Sample Survey*, India Studies in Business and Economics, https://doi.org/10.1007/978-981-96-7320-9_6

42nd round (July 1986–June 1987) till the 64th round (July 2007–June 2008). In this annual series, key information on E-U was also collected from NSS 45th round (July 1989–June 1990).[3] The quinquennial surveys on HCE and E-U got discontinued after NSS 68th round. Instead, the Periodic Labour Force Survey (PLFS) covering the subject of E-U was introduced during 2017–18 and it is continuing as an annual survey. As regards the HCE, the subject got covered twice during the years 2022–23 and 2023–34 (August to July of the respective years) through the Household Consumption Expenditure Survey (HCES).

It is important to mention that in the quinquennial surveys, data on HCE and E-U were collected from different sample households within each FSU since NSS 55th round (July 1999–June 2000) to reduce the canvassing time of schedules in the sample households unlike the practice followed in the previous surveys where the information on both the subjects was collected from the same set of sample households.

The surveys on both subjects adopt a stratified multi-stage design with villages/UFS blocks as the FSUs[4] and households (HH) as the Ultimate Stage Units (USUs). Method of stratification of FSUs/USUs and the sample size of FSUs/number of USUs per FSU differ in the two surveys and the readers may refer to Chapter 3 for the details of sample design including the sample size adopted in some of the above surveys. However, Statement 6.1 describes briefly the method of stratification and sample size as adopted in the latest surveys namely NSS 68th round, HCES and PLFS on these subjects.

For collecting information on various items of HCE, three different reference periods have been used in NSS 68th round and HCES namely (i) last 7 days for edible oil; egg, fish and meat; vegetables; fruits; spices; beverages and processed foods; pan, tobacco and intoxicants; (ii) last 30 days for other food items; fuel and light; miscellaneous goods and services including medical (non-institutional) and rents; and (iii) last 365 days for clothing; bedding; footwear; education; medical (institutional) and durable goods.[5] As regards milk and milk products, the reference period was last 30 days in NSS 68th round and last 7 days in the HCES. Irrespective of the items in terms of either value of expenditure (high or low), frequency of purchase (i.e., frequent or infrequent/occasional) or the length of the reference period, data on all the items of HCE are collected in the NSS through personal interview based on oral enquiry.

In the surveys on E-U conducted by the NSS which includes the PLFS, data on the activity status of the persons in the sample households are collected according to

[3] As regards the sample size, the quinquennial surveys had a larger number of sample households per FSU (about 8 to 10 households) as against a 'thin' sample of 2 to 4 households per FSU in the annual surveys.

[4] In addition to villages/UFS blocks, 'sub-units' have also been treated as the FSUs in the HCES.

[5] The collection of data as per three reference periods for the specified items is popularly known in the NSS as the 'Modified Mixed Reference Period' (MMRP) approach of data collection. Apart from the MMRP approach, NSS 68th round also collected data on different items of HCE based on the Uniform Reference Period (URP) of 30 days and Mixed Reference Period (MRP) of 30 days as well as 365 days. For the details, one can refer to NSS Report Number 555 (MoSPI, 2014b).

Statement 6.1 Comparative sample designs of NSS 68th Round, HCES and PLFS

Survey	Subjects Covered	Stratification		Number of Sample FSUs Allocated		Allotted Sample Size per FSU
		FSUs	USUs	Rural	Urban	
NSS 68th Round	HCE and E-U	Each district is a basic stratum for both rural & urban. In urban, each city with a population of 1 million or above is carved out from a district to treat as a separate stratum by itself	Rural: 3 strata formed, i.e., relatively affluent HHs (str.1); remaining HHs with principal earning from non-agriculture (str.2); and all remaining HHs (str.3) Urban: 3 strata formed as per level of MPCE (top 10% HHs forming str.1, middle 60% as str.2 and bottom 30% as str.3)	7,508	5,276	8 (Allocation of 2 HHs to each of str.1 & str.3 and 4 HHs to str.2)
HCES: 2022–23	HCE	State/UT is the basic stratum for both rural & urban. Towns/cities/blocks of each State/UT are further stratified by population size class of towns x type of UFS bocks (Affluent or not)	3 strata of HHs formed as per the size of land possessed in rural and based on whether or not the HHs owned four-wheeler cars for non-commercial use in case of urban	8,723*	6,115*	18 (This total sample size was allocated over 3 strata in proportion to the number of HHs listed)
PLFS: 2023–24	E-U	Each NSS Region is a stratum in both rural and urban. In urban, towns/cities within each NSS region are stratified based on their population size class	3 strata in rural and 4 strata in urban formed as per the number of members in the household who had completed secondary level of education	7,016	5,784	8

* Number actually surveyed
[See Chapter 3 for more details]

three specified reference periods of (i) the last 365 days preceding the date of survey (known as the usual activity status), (ii) the last 7 days preceding the date of survey (known as the current weekly status) and (iii) each day of the reference week which is known as the current daily status of the person. Data are collected by the field investigators through personal interview as in the surveys on HCE.

6.2 Methodology of Other Selected Countries

6.2.1 Household Expenditure Survey

6.2.1.1 *Australia*[6]

The Household Expenditure Survey (HES) of Australia provides estimates of household expenditure patterns and composition across the population. Previously, the HES was conducted during 1974–75, 1975–76, 1984, 1988–89, 1993–94 and 1998–99. Commencing in 2003–04, the HES has been integrated with the Survey of Income and Housing (SIH) every six years with the year 2015–16 being the latest cycle of the joint surveys. The 2015–16 HES collected information from a sample of approximately 10,046 households over the period July 2015 to June 2016. The document 'Household Expenditure Survey, Australia: Summary of Results methodology, Reference period: 2015–16, Financial year; Released on 13/09/2017' available on the website of the Australian Bureau of Statistics (ABS) gives details about the survey methodology of the HES. Some salient aspects of the survey taken/reproduced from the document are described in the following paragraphs.

The HES sample 'was designed in conjunction with the SIH. In the combined sample, some dwellings were selected to complete both the SIH questionnaire and the HES questionnaire, while other dwellings were selected to complete the SIH questionnaire only. Dwellings were selected through a stratified, multi-stage cluster design from the private dwelling framework of the ABS Population Survey Master Sample. Selections were distributed across a twelve-month enumeration period so that the survey results are representative of income and expenditure patterns across the year'.

The survey collected information by personal interview from usual residents of private dwellings in urban and rural areas of Australia (excluding very remote areas), covering about 97% of the people living in Australia. Residents of non-private dwellings were excluded. Information for each household[7] was collected using:

[6] This section contains information from the Australian Bureau of Statistics (2017).

[7] Australian Bureau of Statistics (2017).

- 'a household level computer assisted interview questionnaire which collected information on household characteristics, expenditure common to all household members (e.g., utility bills), and irregular or infrequent expenditure (e.g., household appliances and holidays overseas)[8];
- an individual level computer assisted interview questionnaire which collected information on income, non-cash benefits, salary sacrifice expenditure and other personal characteristics from each usual resident aged 15 years and over[9]; and
- a personal diary in which usual residents aged 15 years and over recorded their expenditure over two weeks'.[10]

6.2.1.2 *Canada*

The Survey of Household Spending (SHS) undertaken by Statistics Canada is conducted every two years. The target population is the 'population of Canada's 10 provinces, as well as the territorial capitals of Whitehorse, Yellowknife and Iqaluit, excluding residents of institutions, members of the Canadian Forces living in military camps and people living on Indian reserves. In all, these exclusions account for about 2% of the population'.[11] The data are collected on a continuous basis from January to December of the survey reference year. Data gets collected in the form of 12 sub-samples of equal size 'from a sample of households spread over twelve monthly collection cycles'. The sample for the 2021 SHS comprised 39,038 households across the 10 provinces and 1,695 households across Whitehorse and Yellowknife.[12]

The documentation 'Survey of Household Spending (SHS)', as referred in the previous paragraph, gives a detailed description of the survey methodology. The SHS is a sample survey with a cross-sectional design. In the 10 provinces, the final sample in 2021 SHS was obtained from two distinct sampling designs. In the first one, which adopted a two-stage design, the sampling unit at the first stage was the geographical area (called a cluster). In the second stage, the sampling unit was the dwelling. A sample of dwellings was selected from the list of dwellings prepared within each sample cluster. In the second sampling design, the sampling unit was directly the dwelling. In the three territorial capitals, the sampling unit was the dwelling.

The SHS data are collected using both a questionnaire and a diary.[13] The questionnaire is generally used to collect expenditures for more expensive, and less frequently purchased goods and services. The diary is used to collect expenditures for smaller, less valuable items and those purchased more frequently.

---

[8] Ibid.

[9] Ibid.

[10] Ibid.

[11] Statistics Canada (2024a).

[12] Ibid.

[13] This practice is prevalent since 2010 in the provinces and since 2015 in the three territorial capitals.

6.2.1.3 *South Africa*[14]

Statistics South Africa conducts the Income and Expenditure Survey (IES) in the country generally once in every five years with a view to ascertaining the average expenditure patterns of households in different areas of the country. The latest survey relates to the year 2022–23. The IES has a national coverage with the universe comprising private dwellings, workers' hostels, residential hotels and nurses' and doctors' quarters but it excludes student hostels and old-age homes.[15]

As regards the sampling procedure,[16] the sampling frame for the IES 2010–11 was obtained from Statistics South Africa's Master Sample based on the 2001 Population Census enumeration areas (EAs) which were treated as the primary sampling units (PSUs). The IES 2010–11 sample comprised a sample of 3,254 PSUs and 31,419 sampled dwelling units (DUs). In the cases where multiple households were residing at a sampled DU, all the households in the DU were covered.

A household in the IES: 2010–11 was in a sample for a period of four weeks. The required data were collected through a household questionnaire and weekly diaries. The fieldwork for the entire survey was split into 26 survey periods with each revolved around a 'two-week diary-keeping period'. The household questionnaire is split into different modules and the same were administered to respondents during the course of the survey period with one module completed during each visit to the household. As regards the weekly diaries, it is a booklet that was left with the sample households to record the requisite details. A household completed a different diary for each week of the survey period. For the collection of data on household consumption expenditure, three main approaches were used: the acquisition approach, the payment approach and the consumption approach. For further details, the readers can refer to the 2017 publication titled 'Poverty Trends in South Africa: An examination of absolute poverty between 2006 and 2015' by Statistics South Africa which gives a detailed description of the IES: 2005–06 and 2010–11 along with the instruments of data collection. However, the salient features of the IES: 2010–11 are reproduced from the publication in Box 6.1.

6.2.1.4 *The United Kingdom*[17]

The Living Costs and Food Survey (LCF) of the UK is a household survey aimed at collecting information on household expenditure patterns and food consumption. It is a voluntary survey of private households. Each individual in a selected household is asked to complete a household interview and then an expenditure diary for two weeks. The survey is continuous with interviews spread evenly over the year ensuring

[14] This section contains information from Statistics South Africa (2017).

[15] DataFirst (2024).

[16] Ibid.

[17] This section contains information from documents for The Office of National Statistics.

Box 6.1 Salient features of the income and expenditure survey, 2010–11, South Africa

Description	Particulars	Description	Particulars
Reference year	2001	Number of survey periods	26
Sample size (DUs)	31,419	Visits per household	Four
Methodology	Diary and recall	Classification of expenditure items	COICOP*
Household questionnaire	Four modules	Data collection period	23 August 2010 to 4 September 2011
Diaries	Two weekly diaries	Diary-keeping period	30 August 2010 to 29 August 2011
Expenditure data collection approach: Goods = > Acquisition approach; Services = > Payment approach and Own production = > Consumption approach			
Data collection activities by week:			
Week 0 (Week before diary-keeping)	Hand-over by publicity team; Establish rapport with the household; Train household on diary completion; Conduct interview 1		
Weeks 1 to 2 (Diary-keeping weeks)	Drop weekly diaries to be completed by household; Conduct interviews 2 and 3; Collect completed diaries for week 1; Verify completed diaries for week 1; Codification by means of the summary questionnaire		
Week 3 (Week after diary-keeping)	Conduct interview 4; Collect completed diaries for week 2; Verify completed diaries for week 2; Codification by means of the summary questionnaire		

Note *Classification of Individual Consumption according to Purpose
Source Statistics South Africa (2017)

that seasonal effects are covered. The April 2022–March 2023 LCF had a sample size of approximately 4,500 households.[18]

6.2.1.5 *The United States*[19]

The Consumer Expenditure Surveys (CE) in the United States are nationwide household surveys conducted by the U.S. Bureau of Labor Statistics (BLS). The surveys are aimed at obtaining 'information on the complete range of consumers' expenditures, income and demographic characteristics, in the same survey, directly from consumers. BLS publishes 12-month estimates of consumer expenditures annually, with the estimates summarized by various income levels and demographic characteristics'.[20]

---

[18] For more details, readers may see the relevant documents on the subject available on the website of the Office for National Statistics (ONS, 2024a, 2024c).

[19] This section contains information from various sections of U.S. Bureau of Labor Statistics (2022).

[20] U.S. Bureau of Labor Statistics (2022).

The CE comprises two separate surveys namely the Interview Survey and the Diary Survey. While the former is meant for collecting data on large and/or recurring expenditures that respondents are expected to recall, the Diary Survey is intended to collect information on frequently purchased items which can be difficult to recall even a few weeks later.

On behalf of the Bureau of Labor Statistics (BLS), the U.S. Census Bureau collects CE data from a representative sample of households across the United States. The Census Bureau collects data for the Interview Survey as well as the Diary Survey. BLS first draws the primary sampling units (PSUs). Subsequently, the Census Bureau draws a random sample of households inside those geographic areas.

For the sampling of households, first a stratified random sample of PSUs is drawn. The PSUs are clusters of counties grouped together into geographic entities called 'core-based statistical areas' (CBSAs).[21]

Since 2015, the geographic sample in the survey consists of 91 PSUs classified into three categories based on the three PSU types namely (i) Self-representing PSUs being those selected with certainty (23 in number, which are metropolitan CBSAs with a population over 2.5 million people); (ii) Non-self-representing PSUs selected randomly (52 in number, which are metropolitan and micropolitan CBSAs with a population under 2.5 million people) and (iii) Rural PSUs (16 in number, which are non-CBSA areas).

From each of these 91 sample PSUs, a sample of addresses is selected from the list of addresses as per the Census Bureau's master address file (MAF). While 99% of the addresses used in the survey come from the MAF, it is supplemented by a sample list of housing units 'that are owned or managed by organizations for residents who live in group arrangements such as college dormitories and retirement communities'.

'The Interview Survey is a rotating panel survey in which approximately 13,000 addresses are contacted each calendar quarter of the year for the survey. One-fourth of the addresses that are contacted each quarter are new to the survey. Usable interviews are obtained from approximately 5,000 households at those addresses each quarter of the year. After a housing unit has been in the sample for four consecutive quarters, it is dropped from the survey, and a new address is selected to replace it. The Census Bureau selects a sample of approximately 18,000 addresses per year from the MAF and group quarters frames to participate in the Diary Survey (U.S. Bureau of Labor Statistics, 2022).'

---

[21] There are two types of CBSAs: metropolitan and micropolitan. Metropolitan CBSAs are areas that have an urban 'core' of 50,000 or more people, plus the adjacent counties that have a high degree of social and economic integration with the core as measured by commuting ties. Micropolitan CBSAs are similar to metropolitan CBSAs but they have an urban core of 10,000 to 50,000 people. Areas without an urban core or whose urban core is under 10,000 people are called non-CBSA areas. These non-CBSA areas are considered rural PSUs. U.S. Bureau of Labor Statistics (2022).

6.2.2 *Labour Force Survey*

Labour Force Survey (LFS) is another important survey conducted by different countries. Statement 6.2 briefly summarizes some salient features of the survey methodology adopted in this survey by India (through its PLFS) and four other countries namely Australia, Canada, the United Kingdom and the United States. While the survey is conducted on a monthly basis in Australia, Canada and the United States, the periodicity of the survey is quarterly in South Africa and United Kingdom. So far as India is concerned, the quarterly estimates of key labour force parameters are released only for the urban areas of the country and that way the periodicity is quarterly for urban India. However, the periodicity of the survey is an annual one for rural India.

A rotational sampling design is adopted for the selection of samples by each of these countries (for urban areas only in case of India), although the exact scheme of rotation varies across the countries. As regards the data collection, it is important to note that in Australia as well as in the United Kingdom, generally the first interview is conducted in-person while the subsequent ones are through telephone only. In case of the United States, while the data for the first month and fifth month (after an eight-month dormant period) are collected through personal interviews, the other interviews are conducted through telephone only. For the benefit of the readers, the broad survey methodologies adopted for the LFS by Australia, Canada, South Africa, the United Kingdom and the United States are described in the succeeding few paragraphs.

6.2.2.1 *Australia*[22]

The monthly LFS conducted by the Australian Bureau of Statistics (ABS) provides key information about the labour market activity of Australia's resident civilian population aged 15 years and above. The survey is being carried out since 1960. It aims to provide key labour force estimates for the entire country as well as its each state and territory. Initially, it was launched as a quarterly survey. Subsequently, in February 1978, the frequency of the survey was changed from quarterly to monthly. The survey is 'based on a multi-stage area sample of private dwellings. Households within selected dwellings are interviewed each month for eight months, with one-eighth of the sample being replaced each month. Information is obtained either by trained interviewers or through self-completion online. Generally, the first interview is completed face-to-face and subsequent interviews are conducted by phone. All respondents in the sample are also offered the option of completing the survey online. The interviews are generally conducted during the two weeks beginning on the Sunday between the 5th and 11th of each month. The information relates to the

[22] This section contains text from the Australian Bureau of Statistics (2023).

Statement 6.2 Comparison of survey methodology of labour force survey for selected countries

India	Australia	Canada	The United Kingdom	The United States
Name of the Survey				
Periodic Labour Force Survey	Labour Force Survey	Labour Force Survey	Labour Force Survey	Current Population Survey
Survey Organization				
National Sample Survey Office	Australian Bureau of Statistics	Statistics Canada	Office for National Statistics	Bureau of Labour Statistics
Frequency/Periodicity				
Rural—Annual; Urban—Quarterly [Monthly components for both rural and urban introduced in PLFS 2025]	Monthly	Monthly	Quarterly	Monthly
Scope				
Usually, resident, civilians of all ages excluding floating population, orphans and convicted prisoners undergoing sentence	Usually resident, civilians aged 15 years and above	Civilian non-institutionalized population aged 15 years and above	Permanent residents aged 16–74 years	Usually resident, civilian non-institutional populations aged 16 years and above
Sample Size				
Rural—Nearly 56,000 households (243,000 persons) per year; Urban—Approx. 45,000 households (170,000 persons) per quarter^	Approx.26,000 dwellings (52,000 persons)	Approx.56,000 dwellings (100,000 persons)	Approx 40,000 dwellings (100,000 persons)	Approx 60,000 dwellings (112,000 persons)

(continued)

Statement 6.2 (continued)

India	Australia	Canada	The United Kingdom	The United States
Sample Rotation (Rotating Panel Design)				
Rotating design only for urban with 4 independent panels/rotation groups in each quarter. Selected households remain in the survey for four consecutive quarters. A new rotation group/panel is introduced in each quarter to replace an outgoing panel (one-fourth of the sample). For rural, the entire sample for a year is drawn at random^^	Every month has 8 rotation groups. Selected households remain in the survey for eight consecutive months. A new rotation group is introduced each month to replace an outgoing group (one-eighth of the sample)	Each month has 6 rotation groups. Selected households remain in the survey for six consecutive months. A new rotation group is introduced each month to replace an outgoing group (one-sixth of the sample)	Selected households/respondents are questioned five times at 13-week intervals (consecutive) and one-fifth of the sample is replaced each quarter	Each month has 8 rotation groups. Each group is in the sample for 4 consecutive months, leaves the sample during the following 8 months and then returns for another 4 consecutive months. 75% of the sample is common from month-to-month and 50% one year apart for the same month
Method of Data Collection				
Personal interview	Personal interview, telephone interview and online form. Data is collected for each in-scope household member from 'Any Responsible Adult'	Personal or telephone interview. Data was collected from a knowledgeable household respondent	First interview conducted in-person, with subsequent interviews via telephone	Personal interview conducted in first and fifth month (after 8-month dormant period). Other interviews via telephone. Data collected from a responsible adult household

Note: ^Sample size increased in PLFS 2025 (See MoSPI (2025)) ^^ Rotating panel design extended to rural areas also along with monthly rotation scheme in PLFS 2025 (*Source* Relevant information in respect of the countries other than India is as per the documentation in 'Labour force survey—Labour statistics: Concepts, sources and methods; Reference period: 2023' (released on 24/11/2023) by the Australian Bureau of Statistics—accessed from their website on November 8, 2024)

week before the interview (i.e., the reference week)'.[23] A sample of about 24,000 dwellings covering approximately 50,000 people is generally covered.

The sample is in the form of 8 rotation groups, with each of them remaining in the survey for eight months. A new rotation group is introduced each month to replace an outgoing rotation group. A composite estimation method is adopted and the same 'combines the data collected in the previous six months with the current month's data to produce the current month's estimates. This takes advantage of the high correlation between overlapping samples across months in the survey. After composite estimation methods have been applied, the seven months of data are weighted to align with the current month's population benchmarks'.[24] The key estimates are available as original, seasonally adjusted and trend series.

6.2.2.2 *Canada*[25]

The Canadian Labour Force Survey (LFS) is conducted with a monthly frequency. For the details about the survey coverage, questionnaire, data sources and detailed methodology, the readers may like to refer the documentation available under 'Labour Force Survey (LFS), Detailed information for November 2024, Record Number: 3701, Data release—December 6, 2024' on the official website of Statistics Canada.

As the documentation mentions,[26] the target population of the LFS includes 'all persons aged 15 years and older whose usual place of residence is in Canada, including both non-permanent residents (NPRs)—that is, those with a work or study permit, their families, asylum claimants, protected persons and related groups—as well as permanent residents (landed immigrants) and the Canadian-born population. Populations excluded from the LFS target population include those living on reserves, full-time members of the regular Armed Forces and persons living in institutions (including inmates of penal institutions and patients in hospitals and nursing homes)'.

The LFS is a sample survey with a cross-sectional design. The survey adopts a stratified multi-stage design where each province is first divided into 'large geographic strata'. From each stratum, a random sample of first-stage units—which are smaller geographic areas called clusters—is selected. The second-stage units of sampling are dwellings from within each selected cluster.

The LFS adopts a 'rotating panel sample design' where the selected dwellings remain in the sample for six consecutive months. Every month about one-sixth of the sample dwellings are in their first month of the survey, another one-sixth in their second month of the survey, and so on. Within the sample dwellings, while the basic demographic information is collected for all household members, the information on

---

[23] Australian Bureau of Statistics (2024).

[24] Australian Bureau of Statistics (2024).

[25] This section contains information from Statistics Canada (2024b).

[26] Statistics Canada (2024b). Data sources and methodology.

labour force is collected for all civilian household members who are of age 15 years and above.

As per the same documentation,[27] recently 'the monthly LFS sample size has been approximately 68,000 households, resulting in the collection of labour market information for approximately 100,000 individuals. It should be noted that the LFS sample size is subject to change from time to time in order to meet data quality, information or budget requirements … . Data are collected directly from survey respondents. Information about all household members is usually obtained from one knowledgeable household member. Such 'proxy' reporting accounts for approximately 65% of the information collected. Responses to survey questions are captured in one of three ways: by in-person interview; by telephone interviewer; or by self-completed electronic questionnaire'.

6.2.2.3 *South Africa*[28]

Statistics South Africa conducts the Quarterly Labour Force Survey (QLFS) aimed at collecting data on the labour market activities of persons aged 15 years and above who live in South Africa. It is a sample survey of households that collects information from about 30,000 dwelling units in which households reside. The reference period for the survey is one week preceding the date of interview.

The 'Statistical Release P 0211 Quarterly Labour Force Survey Quarter 3: 2024' by Statistics South Africa (November 2024) presents the latest quarterly estimates of key labour force parameters (aggregates as well as rates and ratios) and quarterly changes in the estimates for the country and its provinces. This documentation/statistical release also describes the sampling strategy adopted in the QLFS including the method of estimation. It may be seen that the latest QLFS adopted the 2013 Master Sample. As per the documentation, the Master Sample is 'based on information collected during the 2011 Census. In preparation for Census 2011, the country was divided into 103,576 enumeration areas (EAs). Census EAs, together with auxiliary information for the EAs, were used as frame units or building blocks for formation of primary sampling units (PSUs) for the Master Sample. They covered the entire country and had other information that is crucial for stratification and creation of PSUs. There are 3,324 primary sampling units (PSUs) in the Master Sample, with an expected sample of approximately 33,000 dwelling units (DUs)….The Master Sample is designed to be representative at provincial level and at metro/non-metro levels. Within metros, the sample is further distributed by geographical type. The three geography types used are Urban, Tribal and Farms. A sample for the QLFS is based on a stratified two-stage design with probability proportional to size (PPS) sampling of PSUs in the first stage, and sampling of dwelling units (DUs) with systematic sampling in the second stage'.

[27] See the sections on Sampling and Data sources.

[28] This section contains information from Statistics South Africa (2024).

The QLFS adopts a rotational sample design. In each quarter, a quarter of the sampled dwellings are rotated out of the sample and they are 'replaced by new dwellings from the same PSU or the next PSU on the list. Thus, sampled dwellings are expected to remain in the sample for four consecutive quarters'. As the documentation mentions, 'the sampling unit is the dwelling, and the unit of observation is the household. Therefore, if a household moves out of a dwelling after being in the sample for two quarters (for example) and a new household moves in, the new household will be enumerated for the next two quarters. If no household moves into the sampled dwelling, the dwelling will be classified as vacant (or unoccupied)'.

As regards the estimates and variance estimation, according to the said documentation: 'final survey weights are used to obtain the estimates for various domains of interest, e.g., number of persons employed in Agriculture in Western Cape, the number of females employed in Manufacturing, etc. ... The Fay's BRR[29] method has been used for variance estimation in the QLFS because of its simplicity'.

6.2.2.4 *The United Kingdom*[30]

The Labour Force Survey (LFS) of the UK is a quarterly survey. It adopts a rotational sampling design and the households in the sample are retained for a total of five consecutive quarters (known as five waves). The interviews are scheduled to take place exactly after 13 weeks so that the fifth interview takes place just after one year from the first interview.

The document '*Labour Force Survey—user guidance, 2024 Update, Volume I: Background and methodology 2024*' available on the official website of the Office for National Statistics gives a detailed description of the survey methodology adopted in the LFS. To cite from this document, 'same number of Wave 1 (new) addresses are selected each quarter. So, in any given quarter, about one-fifth of the addresses in the entire sample are in Wave 1, one-fifth in Wave 2, and so on. Thus, between any two consecutive quarters, about 80% of the selected addresses are in common....Four different sampling frames are used in the UK Labour Force Survey. Great Britain is split into two areas: south of the Caledonian Canal, comprising all of England, Wales and most of Scotland; and north of the Caledonian Canal in Scotland. Northern Ireland has its own sampling frame..... The Wave 1 sample is selected by first ordering the sampling frames geographically, and then drawing the selection systematically (that is, with a fixed interval). The subsequent waves are not drawn from the frames; the Wave 1 selections are simply retained and become Wave 2 interviews in the next quarter and so on. For the most part, the LFS may be regarded as a single-stage sample of households each quarter'.[31] From the Q3 2010 survey, only one household

[29] Balanced Repeated Replication.

[30] This section contains information from various documents of Office for National Statistics, UK.

[31] ONS (2024b). With certain changes introduced in 2010, this is strictly no longer the case now. For the changes made, the interested readers may refer to Sect. 3.5 of the document.

was selected for interview if there was more than one present at the sampled address. However, earlier all households at such an address were interviewed.

The fieldwork is carried out by both face-to-face interview and by telephone interview (see the document referred above for further details). The major part of the LFS questionnaire seeks information about the respondents' activities during a 7-day period which ends on a Sunday. The sample data is duly weighted to arrive at the final estimates. The LFS uses 'calibration weighting' to assign a 'calibration weight' to each responding individual. 'These calibration weights are set to sum to a set of calibration totals within calibration groups—for example, the weights of all 18-year-old males in an LFS dataset (a calibration group) will sum to the population total of eligible 18-year-old males in the UK (a calibration total) at the time the survey was taken[32]'.

6.2.2.5 *The United* States[33]

The Current Population Survey (CPS), being the source of labour force statistics, adopts a stratified multi-stage design with a county or a group of contiguous counties comprising the primary sampling units (PSUs) and households as the ultimate stage sampling units. For details of the sample design adopted in the CPS, the readers may see the documentation 'Current Population Survey: Design' available on the website link: Bureau of Labor Statistics > Publications > Handbook of Methods. The salient features of the sample design taken/partly reproduced from this documentation are summarized in the succeeding few paragraphs.

The whole geographical area of the United States is divided into 1,987 PSUs. Metropolitan areas within a state 'are used as a basis for forming many PSUs. Outside of metropolitan areas, two or more counties normally are combined to form a PSU … . The 1,987 PSUs are grouped into strata within each state. Then, one PSU is selected from each stratum, with the probability of its selection proportional to the population of the PSU. In the first stage of sampling, 852 sample areas are chosen from the strata. Nationally, there are 506 PSUs in strata by themselves. These strata are self-representing and generally are the most populous PSUs in each state. The remaining 346 strata are formed by combining PSUs that are similar in such characteristics as unemployment, the proportion of housing units with three or more people, the number of people employed in various industries, and average monthly wages for various industries'.

The sample design in the CPS is state-based and the sampling ratio for the purpose of selecting households depends on 'the size of the state's population, as well as both national and state reliability requirements. The state sampling ratios range roughly from 1 in every 200 households to 1 in every 3,000 households'. As mentioned in the above-cited documentation, the current 'within-PSU sample design was developed with the use of block-level data from the 2010 census. Normally, census blocks are

[32] ONS (2019, 2024b). For more details, the readers may refer to these documents.

[33] This section contains information from the U.S. Bureau of Labor Statistics (2018).

bounded by streets and other prominent physical features, such as rivers or railroad tracks. County, minor civil division and census place limits also serve as block boundaries. In cities, blocks can be bounded by four streets and be quite small in land area. In rural areas, blocks can be several square miles in size. Within each block, housing units are sorted geographically and grouped into clusters. A systematic sample of these clusters is then independently selected from each stratum by using the appropriate within-PSU sampling ratio.…The CPS starts with a probability sample of about 74,000 assigned housing units each month. Approximately 62,000 housing units are eligible for interview each month …. Interviews are completed for about 54,000 housing units'.

For the purpose of estimation, after the completion of second-stage (SS) weighting and by making use of the 4–8-4 rotation pattern adopted in the CPS, the composite (CMP) estimator of the current period/month 't' is derived as per the following expression[34]:

$$\hat{Y}_t^{CMP} = (1 - K)\hat{Y}_t^{SS} + K\left(\hat{Y}_{t-1}^{CMP} + \Delta_t\right) + A\,\hat{\beta}_t$$

Where $\hat{Y}_t^{SS} = \sum_{i=1}^{8} x_{ti}$

$\Delta_t = \frac{4}{3}\left[\sum_{i=2}^{4}\left(x_{ti} - x_{t-1,i-1}\right) + \sum_{i=6}^{8}\left(x_{ti} - x_{t-1,i-1}\right)\right]$

$\hat{\beta}_t = (x_{t1} + x_{t5}) - \frac{1}{3}\left(\sum_{i=2}^{4} x_{ti} + \sum_{i=6}^{8} x_{ti}\right)$

i = MIS (month-in-sample) 1, 2, …, 8

x_{ti} = sum of second-stage weights of respondents in month t and MIS i with characteristic of interest.

K = 0.4 for unemployed, 0.7 for employed,

A = 0.3 for unemployed, 0.4 for employed.

K determines the weight to get the weighted average of two estimators for the current month: (1) the current month's second-stage estimate and (2) the sum of the previous month's composite estimator and an estimator of the change since the previous month. The estimate of change is based on data from sample households in the six rotation groups common to months t and t-1 (about 75%).

'A' determines the weight of β_t, a term adjusting for the difference between the continuing months in sample and the new months in sample.

[34] United States Census Bureau (2019).

6.3 Major Differences in the Approach Adopted by India Vis-à-Vis the Other Countries and Way Forward

From the foregoing discussion, it is apparent that so far as the collection of data on HCE is concerned, the NSS has adopted the personal interview method (oral enquiry) to record household consumption on all the items irrespective of their monetary value, i.e., high or low and the frequency of purchase. On the country, the selected countries under reference have adopted the interview method only in respect of the items with irregular/infrequent/high-value purchases and a diary-based approach for frequently purchased and low-value items which may be difficult to recall even after a few days/weeks. As regards the reference periods for the collection of data on HCE so far as India is concerned, the readers may like to refer to the discussions on NSS 55th Round and NSS 68th Round in Chapter 3.

As regards the labour force surveys, while the periodicity of these surveys varies in different countries,[35] there are some differences in the age group of the target population. The age group is 15 years and above in respect of Australia, Canada and South Africa while the UK and the US consider 16 years and older for the purpose of data collection. In India, the data on E-U are collected without any such minimum threshold on age.[36] All the countries adopt a reference period of the week for data collection.[37] However, in addition, India also adopts a longer reference period of last 365 days to record 'usual activity status' of the persons. For the data collection, in-person interview and/or telephone interview is adopted by different countries.[38] For the labour force surveys, all the selected countries are adopting a rotational sampling design with some re-visits to the same households as is the case with the urban component of the PLFS in India although the number of re-visits to a household and the method of rotation varies over the countries (see Statement 6.2). For producing statistics, while India has been publishing only the rates and ratios[39] like LFPR, WPR and UR, other countries under reference also produce the estimates of aggregates, i.e., number of employed/unemployed/out of labour force and changes

[35] It is monthly in case of Australia, Canada and the US while the periodicity is quarterly in South Africa and the UK. In case of India, the periodicity is annual for rural areas and quarterly for urban till December 2024 and monthly thereafter for both rural and urban.

[36] However, children in the age group of 0–4 years are classified under 'out of labour force' and a very insignificant proportion of children in the age group of 5–9 years may get classified under the labour force.

[37] Mostly, week before the interview; 7-day period ending on a Sunday in case of the UK; the week beginning on a Sunday and ending on a Saturday in case of the US.

[38] For example, in Australia, generally the first interview is completed face-to-face while the subsequent interviews are conducted by phone. Responses to survey questions in Canada are captured in one of the three ways: by in-person interview; by telephone interviewer; or by self-completed electronic questionnaire (see Statement 6.2 for further details).

[39] This compels the data users to derive their own estimates of aggregates like the number of employed, number of unemployed and the number in the labour force under certain assumptions about population projections which quite often leads to different sets of figures by different scholars.

in these aggregates—monthly/quarterly/annual as relevant—in terms of number of persons, in addition to the said rates and ratios.

As a way forward, it may be useful if India, like the other countries, also explores a diary-based method of data collection for more frequently purchased items and low-value items in the surveys on HCE[40] to tackle the likely problem of recall error in the collection of data based on oral enquiry. As regards the surveys on E-U/PLFS, it will be beneficial for the data users if apart from the rates and ratios like the LFPR, WPR and UR, estimates of aggregates, i.e., total number of employed, unemployed and out of labour force are also produced by India as some of the other countries are doing. Further, for the generation of the quarterly estimates based on the PLFS (including the monthly estimates since PLFS 2025) where the sample FSUs have been drawn by rotational sampling design, it is worth considering the generation of 'composite estimates' of LFPR, WPR and UR by also taking into consideration the data of 're-visit' samples introduced during the previous quarters/months. Such a method of composite estimation, as also adopted by some other countries like Australia and the United States, is likely to enhance the level of precision of the estimates of LFPR, WPR and UR. Some other aspects of these two subjects of enquiry requiring attention are also discussed in Chapter 9.

References

Australian Bureau of Statistics. (2017). *Household expenditure survey, Australia: Summary of results methodology, reference period: 2015–16, financial year* (Released on 13/09/2017).

Australian Bureau of Statistics. (2023). *Labour force survey—Labour statistics: Concepts, sources and methods, reference period: 2023* (Released on 24/11/2023).

Australian Bureau of Statistics. (2024). *Labour force, Australia methodology, reference period: September 2024.*

DataFirst. (2024). Report generated on: May 22, 2024; Accessed from their website on January 15, 2025.

MoSPI. (2013). *National seminar on the results of NSS 66th round survey.* National sample survey office, Ministry of Statistics and Programme Implementation, Government of India; accessible at https://mospi.gov.in/sites/default/files/publication_reports/NSS_66th_round_Compenndium_Oct_2017.pdf

MoSPI. (2014a). *Employment and unemployment situation in India 20ll-12, NSS 68th round, July 2011–June 2012, Report number 554.* National Sample Survey Office, Ministry of Statistics and Programme Implementation, Government of India.

MoSPI. (2014b). *Level and pattern of consumer expenditure 20ll-12, NSS 68th round, July 2011–June 2012, Report number 555.* National Sample Survey Office, Ministry of Statistics and Programme Implementation, Government of India.

MoSPI. (2024a). *Annual report, Periodic labour force survey (PLFS), July 2023–June 2024.* National Sample Survey Office, Ministry of Statistics and Programme Implementation, Government of India.

MoSPI. (2024b). *Survey on household consumption expenditure: 2022–23 (HCES: 2022–23), August 2022–July 2023, Report number 591.* National Sample Survey Office, Ministry of Statistics and Programme Implementation, Government of India.

[40] Subject to the feasibility of the method which may be tested based on a pilot exercise.

MoSPI. (2025). *Periodic Labour Force Survey (PLFS), Changes in 2025.* National Statistics Office, Ministry of Statistics and Programme Implementation, Government of India.
ONS. (2019). *Internet users QMI.* Office for national statistics. Newport, UK. Last revised January 18, 2019 (accessed on April 11, 2025). https://www.ons.gov.uk/businessindustryandtrade/itandinternetindustry/methodologies/internetusersqmi
ONS. (2024a). *Family spending in the UK: April 2022 to March 2023.* ONS website, statistical bulletin, Office for National Statistics. Newport, UK.
ONS. (2024b). *Labour force survey—User guidance, 2024 Update, Volume I: Background and methodology 2024.* Office for National Statistics. Newport, UK.
ONS. (2024c). *Living costs and food survey technical report: Financial year ending March 2023.* Office for National Statistics. Newport, UK.
Statistics Canada. (2024a). *Survey of household spending (SHS), definitions, data sources and methods used* (Accessed on 18 November, 2024).
Statistics Canada. (2024b). *Labour force survey (LFS), detailed information for November 2024, record number: 3701, Data release—December 6, 2024.*
Statistics South Africa. (2017). *Poverty trends in South Africa: An examination of absolute poverty between 2006 and 2015.*
Statistics South Africa. (2024). Statistical release P0211, Quarterly labour force survey. *Quarter, 3,* 2024.
U.S. Bureau of Labor Statistics. (2018). *Handbook of methods, Current population survey, Design.* Last modified on April 10, 2018; accessed on June 3, 2025. https://www.bls.gov.opub/hom/cps/design.htm
U.S. Bureau of Labor Statistics. (2022). *Handbook of methods; Consumer expenditures and income.* Statistics, Publications. Last modified on September 12, 2022; accessed on April 11, 2025; https://www.bls.gov/opub/hom/cex/home.htm
United States Census Bureau. (2019). *Design and methodology, current population survey, October 2019, Technical paper 77.*

Chapter 7
NSS Design-Based Estimates and Multiplier

As discussed in Chapters 2 and 3, the NSS adopts a stratified multi-stage design with villages /urban blocks (and 'sub-units' in some recent surveys) as the FSUs and households/enterprises as the USUs. In the large FSUs, HGs/SBs serve as an intermediate stage of sampling.[1] Based on the sample, inference about the population parameters is drawn by deriving the estimates after adjusting the sample data with appropriate weights—popularly known as 'multipliers'—as per the sample design adopted in the survey. The survey estimates are generally published in the form of either aggregates or ratios or in the form of percentage distribution of the target population according to some characteristics of interest. The subsequent portion of this chapter describes how the survey-based estimates are actually derived and the concept of multiplier (see also MoSPI 2014, 2018, 2019, Murthy 1967 and Murthy and Roy 1975).

7.1 The Multiplier

The multiplier is a conversion (i.e., multiplying) factor by which the sample data is inflated or multiplied to infer about the universe from which the sample is drawn. The formula for the multiplier actually depends upon the sampling procedure based on which the sample has been selected. To give a simpler example, suppose from a population of N enterprises, a random (SRSWOR) sample of n enterprises is drawn to estimate how many workers may be working in the universe of N enterprises. If a total number of k workers are found to be working in the sample of n enterprises, then the estimate ($\hat{Y}$) for the total number of workers in the universe will be:

[1] FSUs are first-stage units and USUs are Ultimate Stage Units. In NSSO (National Sample Survey Office) surveys, HGs stand for Hamlet Groups and SBs stand for Sub-Blocks.

G C Manna, *75 Years of the Indian National Sample Survey*, India Studies in Business and Economics, https://doi.org/10.1007/978-981-96-7320-9_7

$$\hat{Y} = \frac{N}{n} \cdot k \tag{7.1}$$

In the Eq. (7.1), the factor '*N/n*' is treated as the multiplier.

In the above example, if the study variable, i.e., the number of workers in a sample enterprise is denoted by *y*, then the estimated total number of workers in the universe of *N* enterprises can be written mathematically in a more general form as:

$$\hat{Y} = \frac{N}{n} \sum\nolimits_{i=1}^{n} y_i \tag{7.2}$$

where y_i denotes the number of workers in the *i*-th sample enterprise ($i = 1, 2, \ldots., n$) and the factor '*N/n*' by which each y_i value is to be multiplied and then summed up over all sample enterprises is called the multiplier.

It is important to remember that the multipliers are applied to the sample data of each USU (i.e., household/enterprise) for the estimation of aggregates and not ratios. The ratios need to be estimated at the last stage for the domain of interest (like district/State, etc.) as the ratio of domain level estimates of concerned two aggregates corresponding to the numerator and denominator associated with the ratio.

7.2 Design-Based Estimates

7.2.1 Estimates of Aggregates and the Associated Multipliers

7.2.1.1 Case-I: FSUs Selected by PPS with no HG/SB Formation in Sample FSUs

Now let us take the example of a stratified two-stage design with villages/UFS[2] blocks/sub-units as the FSUs (forgetting for the time being that there is any HG/SB selection within the FSU) and households/enterprises as the USUs with proper stratification of USUs as is the case with the NSS. Let us also assume that from each first-stage stratum sample FSUs have been selected by PPSWR[3] (or even by circular systematic sampling with PPS as was the case in many NSS rounds earlier). Then the estimate of total/aggregate of the study variable for a given first-stage stratum is given by:

$$\hat{Y}_s = \frac{1}{n_s} \sum_{i=1}^{n_s} \frac{y_{si}}{p_{si}} \tag{7.3}$$

[2] Urban Frame Survey.

[3] Probability Proportional to Size With Replacement (PPSWR) sampling.

In Eq. (7.3), s = first-stage stratum, n = number of surveyed FSUs, y = study variable of interest, i = sample FSU, p = probability of selection of the FSU. If FSUs have been selected with 'f' as the size measure of sampling of an FSU (which is generally the census population in case of villages and number of households in respect of UFS blocks in many surveys on households) with 'F' as the total size of the stratum (meaning thereby that $p = f/F$), the above equation gets converted to:

$$\hat{Y}_s = \frac{F_s}{n_s} \sum_{i=1}^{n_s} \frac{y_{si}}{f_{si}} \tag{7.4}$$

Values of the study variable for the FSU as a whole, i.e., y_{si} in Eq. (7.4) are not known since information on the study variable for each SSU of the FSU is unknown. This problem is tackled by introducing another stage of sampling, i.e., (a) by listing USUs within the FSU, (b) by grouping the USUs into a few second-stage strata and (c) by finally selecting a random sample of USUs from each second-stage stratum (SSS) by SRSWOR[4] (or systematic sampling with equal probability) and interviewing them to ascertain the value of the study variable at the level of USU. Under such a situation, y_{si} is estimated as under:

$$y_{si} = \sum_{j=1}^{J} y_{sij}$$

$$\text{i.e.,}\ y_{si} = \sum_{j=1}^{J} \frac{H_{sij}}{h_{sij}} \sum_{k=1}^{h_{sij}} y_{sijk} \tag{7.5}$$

where j = SSS, J = total number of second-stage strata formed within the FSU, H = total number of USUs listed, h = number of USUs surveyed, k = sample USU and y = value of the study variable of interest for the selected SSU.

With the expression of y_{si} from Eq. (7.5) posted in Eq. (7.4), the formula for an estimate of the aggregate for s-th first-stage stratum is given by:

$$\hat{Y}_s = \frac{F_s}{n_s} \sum_{i=1}^{n_s} \frac{1}{f_{si}} \sum_{j=1}^{J} \frac{H_{sij}}{h_{sij}} \sum_{k=1}^{h_{sij}} y_{sijk} \tag{7.6}$$

The 'multiplier' (M) attached to k-th sample USU (i.e., household/enterprise) is nothing but the mathematical expression without summation signs in the above formula (i.e., the product of three ratios) and it is given by:

$$M_{sij} = \frac{F_s}{n_s} \cdot \frac{1}{f_{si}} \cdot \frac{H_{sij}}{h_{sij}} \tag{7.7}$$

[4] Simple Random Sampling Without Replacement (SRSWOR).

It is apparent from Eq. (7.7) that the value of the multiplier attached to a sample USU will not only change over the first-stage strata and over the FSUs within a first-stage stratum but it will also differ across USUs belonging to different second-stage strata within a given FSU since the values of H_{sij} and quite often the values of h_{sij} also will vary over the second-stage strata within an FSU. It is also important to note that the value of M_{sij} will be exactly the same for all the USUs within a given SSS of the FSU.

In the NSS, sample FSUs from each first-stage stratum are generally selected in the form of two independent sub-samples having the same number of sample FSUs in each sub-sample. Accordingly, the formula in Eq. (7.6) is to be applied for each sub-sample to first derive sub-sample-wise estimates of aggregates for s-th stratum where F_s will be the same for each sub-sample and n_s will represent a total number of FSUs surveyed under a given sub-sample. Thereafter, the pooled estimate of $\hat{Y}_s$ based on the two sub-samples will be obtained as the simple average of two sub-sample wise estimates of aggregate for the s-th stratum. Similarly, in Eq. (7.7), n_s will represent the number of surveyed FSUs under a given sub-sample of s-th stratum.

In many NSS rounds, each stratum is divided further into a certain number of sub-strata and sample FSUs are drawn from each sub-stratum. Under these situations, in the above expressions, the symbol 's' will represent the sub-stratum.

Once stratum/sub-stratum level estimates of aggregates are obtained, the addition of these estimates over all the strata/sub-strata within a State, separately for rural and urban areas, will produce aggregate the estimate for the State (rural/urban). The addition of State/UT-wise estimates will give the aggregate estimate for all-India.

The procedure discussed above to derive State/all-India level estimates of aggregates using the stratum/sub-stratum level estimates will apply to the other cases also described in the succeeding paragraphs.

7.2.1.2 Case-II: FSUs Selected by SRSWOR with no HG/SB Formation in Sample FSUs

In this case (i.e., FSUs selected by SRSWOR), F_s in Eqs. (7.6) and (7.7) for respectively deriving aggregates and multipliers gets replaced with N_s representing the total number of FSUs in the frame of s-th first-stage stratum and the expression, i.e., ratio $1/f_{si}$ disappears in both the formulae since $f_{si} = 1$ for each sample FSU. This procedure of replacing F_s with N_s and taking f_{si} values as 1 will apply to the subsequent cases also where the discussion is confined to the PPS method of sampling of FSUs.

7.2.1.3 Case-III: FSUs Selected by PPS with HG/SB Formation/ Selection in Sample FSUs

- Situation 1: Selecting one segment, i.e., HG/SB at random

As per this practice, each FSU is divided into a certain number of HGs/SBs (say, D) out of which one segment, i.e., HG/SB is selected at random (with equal probability). This was the practice adopted in many initial rounds of the NSS to reduce the workload of listing USUs. In such a situation, the right-hand side expression in Eq. (7.5) will be multiplied by D_{si} (with D_{si} denoting the total number of HGs/SBs formed in the FSU) to derive y_{si} and finally the formula for s-th stratum aggregate corresponding to Eq. (7.6) gets revised as follows:

$$\hat{Y}_s = \frac{F_s}{n_s}\sum_{i=1}^{n_s}\frac{D_{si}}{f_{si}}\sum_{j=1}^{J}\frac{H_{sij}}{h_{sij}}\sum_{k=1}^{h_{sij}} y_{sijk} \tag{7.8}$$

Accordingly, the multiplier at the SSU level stands revised as:

$$M_{sij} = \frac{F_s}{n_s}\cdot\frac{D_{si}}{f_{si}}\cdot\frac{H_{sij}}{h_{sij}} \tag{7.9}$$

- Situation 2: Selection of one HG/SB with certainty and another one at random from the remaining HGs/SBs

As mentioned in Chapters 4 and 5, in many NSS rounds/recent surveys (see, e.g., NSS 68th round and the PLFS), out of all the HGs/SBs formed (say, D) within an FSU, the one having a maximum population (say, segment 1) is selected with certainty, i.e., probability '1' and another one HG/SB is selected at random (with equal probability, say, segment 2) from among the remaining (D-1) HGs/SBs. In this case, y_{si}, i.e., FSU level total for the aggregate corresponding to Eq. (7.5) will be derived as:

$$y_{si} = y_{si1} + (D_{si} - 1)y_{si2}$$

where y_{si1} and y_{si2} respectively denotes the contribution of segment 1 and segment 2 in the total for the i-th FSU of s-th stratum. Values of y_{si1} and y_{si2} are derived exactly as per the right-hand side expression in Eq. (7.5) with H and h respectively representing the number of USUs listed and surveyed in the respective segment 1 or 2 as applicable. Accordingly, the formula for the stratum level estimate of aggregate can be written as under where suffix 1 and suffix 2 denote segment 1 and segment 2 respectively:

$\hat{Y}_s = \frac{F_s}{n_s}\sum_{i=1}^{n_s}\frac{1}{f_{si}}\left[y_{si1} + (D_{si} - 1)y_{si2}\right]$ or equivalently,

$$\hat{Y}_s = \frac{F_s}{n_s}\sum_{i=1}^{n_s}\frac{1}{f_{si}}\left[\sum_{j=1}^{J}\frac{H_{si1j}}{h_{si1j}}\sum_{k=1}^{h_{si1j}} y_{si1jk} + (D_{si} - 1)\sum_{j=1}^{J}\frac{H_{si2j}}{h_{si2j}}\sum_{k=1}^{h_{si2j}} y_{si2jk}\right] \tag{7.10}$$

In this procedure, the multiplier attached to a USU for j-th SSS of i-th FSU is derived as per the following formula:

$$M_{si1j} = \frac{F_s}{n_s} \cdot \frac{1}{f_{si}} \cdot \frac{H_{si1j}}{h_{si1j}} (\textit{for each of the USUs belonging to segment } 1) \quad (7.11)$$

$$M_{si2j} = \frac{F_s}{n_s} \cdot \frac{(D_{si} - 1)}{f_{si}} \cdot \frac{H_{si2j}}{h_{si2j}} (\textit{for each USU belonging to segment } 2) \quad (7.12)$$

Thus, $\hat{Y}_s$ in Eq. (7.10) can be written as:

$$\hat{Y}_s = \sum_{i=1}^{n_s} \sum_{j=1}^{J} \sum_{k=1}^{h_{si1j}} M_{si1j} y_{si1jk} + \sum_{i=1}^{n_s} \sum_{j=1}^{J} \sum_{k=1}^{h_{si2j}} M_{si2j} y_{si2jk} \quad (7.13)$$

From Eq. (7.13), it can be seen that the aggregate estimate for the s-th first-stage stratum is obtainable by multiplying the USU (i.e., household or enterprise) level data of the study variable with the corresponding USU-level multiplier[5] and summing these products over all the sample/surveyed households in the stratum. Finally, the State-level aggregate estimate $\left(\hat{Y}\right)$ can be obtained by adding the stratum level aggregates $\hat{Y}_s$ for all the first-stage strata formed within a State or UT,[6] separately for rural and urban areas.

The situation 2 above, which is applicable to both household surveys and enterprise surveys, can be extended to a situation where apart from selecting one HG/SB with certainty (i.e., segment 1), a few more (say, d) HGs/SBs are selected at random (by SRSWOR) out of the remaining (D-1) HGs/SBs formed within the FSU and a combined listing is done for the 'd' randomly selected HGs/SBs (treated as segment 2).[7] In this case, the formula for $\hat{Y}_s$ will remain the same as in Eq. (7.10) with the following changes:

(a) the expression $(D_{si} - 1)$ will be replaced with $\frac{(D_{si}-1)}{d_{si}}$; and
(b) H_{si2j} and h_{si2j} will represent total number of USUs listed and surveyed respectively in the j-th SSS of segment 2 comprising 'd' randomly selected HGs/SBs.

7.2.2 *Estimates of Ratios*

Estimates of ratios like worker population ratio (WPR), household monthly per capita expenditure (MPCE), gross value added (GVA) per worker, etc. are derived for the

[5] The USU-level multipliers are generally posted in the unit-level database of the NSS.

[6] Union Territory.

[7] This type of procedure was followed in the last few enterprise surveys of the NSS including NSS 67th round, NSS 73rd round and ASUSE.

domain of interest, which could be district, State, etc. (and not for each USU) as follows:

$\hat{R} = \frac{\hat{Y}}{\hat{X}}$ where $\hat{R}$, $\hat{Y}$ and $\hat{X}$ respectively denotes the estimate of ratio, numerator corresponding to the ratio (e.g., estimated number of workers in case of WPR) and denominator corresponding to the ratio (i.e., estimated population in case of WPR) for the domain.

7.3 Elaborations

Let us consider the example of Situation 2 under Case-III where the FSU is selected by PPSWR and there has been HG/SB formation in the FSU with one HG/SB selected with certainty (segment 1) and another one (segment 2) selected at random from among the remaining (D-1) HGs/SBs. Suppose the domain of interest is State (rural or urban). If one is interested in estimating total number of households, then the study variable is 'number of households'. Accordingly, the values of y_{si1jk} and y_{si2jk} in Eq. (7.13) for all sample households will be '1'. Thus, the estimated number of households for a stratum, i.e., $\hat{Y}_s$ is the sum of household level multipliers for all the surveyed households in the stratum since all the y-values are equal to 1. And the estimated number of households for the State is the sum of household level multipliers for all the surveyed households in the State. The same principle holds good for estimating a total number of enterprises in the enterprise surveys with the only change that enterprise will be the USU in place of household.

In case one is interested in estimating WPR (which is a ratio), then $\hat{Y}$ and $\hat{X}$ will respectively denote the estimated number of workers and estimated population in the State (rural or urban or rural plus urban). In this case, for deriving $\hat{Y}_s$, and subsequently the State total $\hat{Y}$, values of y in Eq. (7.13) will be the number of workers in the respective surveyed households. Thus, the household level data on a number of workers may be multiplied by the household level multipliers and the summation of these products over all the surveyed households in the State will yield an estimated number of workers in the State. Similarly, for estimating total population $\hat{X}$, y-values will represent the respective household sizes (i.e., number of members in the household).

If the parameter of interest is MPCE,[8] y-values for estimating $\hat{Y}$ will represent household level total consumer expenditure and y-values in the estimation of $\hat{X}$ will represent household sizes. Thereafter, the household average MPCE for the State $(\hat{R})$ will be obtained as $\hat{Y}/\hat{X}$. For estimating GVA[9] per worker in the enterprise surveys, y-values in the estimation of $\hat{Y}$ will represent the GVA of the surveyed enterprises and y-values in estimating $\hat{X}$ will represent the number of workers in the respective enterprises surveyed.

[8] Monthly Per Capita Consumption Expenditure.

[9] Gross value Added.

References

Murthy, M. N. (1967). *Sampling theory and methods*. Statistical Publishing Society.

Murthy, M. N., & Roy, A. S. (1975). Development of the sample design of the Indian National Sample Survey during its first twenty-five rounds, *Sankhya, 37* (Series C, Pt. 1), 1–42.

MoSPI. (2014). *Level and pattern of consumer expenditure 20ll–12, NSS 68th round, July 2011–June 2012, Report number 555*. National Sample Survey Office, Ministry of Statistics and Programme Implementation, Government of India.

MoSPI. (2018). *Economic characteristics of unincorporated non-agricultural enterprises (excluding construction) in India, Report number 582, NSS 73rd round, July 2015–June 2016*. National Sample Survey Office, Ministry of Statistics and Programme Implementation, Government of India.

MoSPI. (2019). *Annual Report, Periodic labour force survey (PLFS), July 2017–June 2018*, National Statistical Office, Ministry of Statistics and Programme Implementation, Government of India.

Chapter 8
Precision of Key Estimates

The level of precision of estimates is generally measured by computing the relative standard error (RSE) of the estimates. The RSEs of estimates of aggregates and ratios in the NSS are usually estimated based on the sub-sample-wise estimates of aggregates at the stratum level.[1] If the value of estimated RSE is within 5%, the estimates are considered to be quite robust signifying that the deviation of the true value in the population from the estimate is a maximum of 10% with a very high probability of at least 0.95. Even if the RSE lies between 5 and 10%, the estimates may be considered as fairly reliable in the sense that the true value, i.e., population parameter is not likely to deviate from the estimate by more than 20%. The subsequent portion of the text in this chapter describes how the RSEs are estimated in the NSS and what is the level of precision of some key NSS estimates as per the latest surveys.

8.1 Method of Estimating Precision/RSE

8.1.1 RSE of Estimates of Aggregates

Chapter 7 provides the formula for estimating stratum (or stratum x sub-stratum if sub-strata formed within the stratum) level aggregates $\hat{Y}_s$. The formula is applicable for each of the two sub-samples. Based on these stratum level aggregates by sub-sample, RSE of estimates of aggregate $\hat{Y}$ for the domain of interest (say, State/UT x rural/urban) in percentage term is generally estimated as under[2]:

[1] Henceforth, the term 'stratum' means stratum x sub-stratum in case sub-strata of FSUs have been formed within a stratum for any given NSS round.

[2] For further details, refer to Sample design and estimation procedure in the Appendix of NSS reports.

G C Manna, *75 Years of the Indian National Sample Survey*, India Studies in Business and Economics, https://doi.org/10.1007/978-981-96-7320-9_8

$$\widehat{RSE}(\hat{Y}) = \frac{\sqrt{\widehat{Var}(\hat{Y})}}{\hat{Y}} X\,100$$

where $\widehat{Var}(\hat{Y}) = \frac{1}{4}\sum_{s}(\hat{Y}_{s1} - \hat{Y}_{s2})^2$ with suffix 1 and suffix 2 denoting sub-samples 1 and 2 respectively.

8.1.2 *RSE of Estimates of Ratios*

Any ratio (R) is a ratio of two aggregates—one (say, Y) coming as the numerator of the ratio and the other (say, X) as its denominator. For example, if R is the average MPCE for the universe like State x rural/urban/combined and if State is the domain of interest, then Y signifies total consumption expenditure of all households in the universe and X represents total number of persons in the universe. Similarly, if R is the unemployment rate, Y and X would respectively mean total number of unemployed and total number of persons in the labour force for the domain of interest.

Estimates of RSE of ratios in the NSS are generally computed as follows (in percentages):

$\widehat{RSE}(\hat{R}) = \frac{\sqrt{\widehat{MSE}(\hat{R})}}{\hat{R}} \times 100$ (with MSE denoting mean square error) where $\widehat{MSE}(\hat{R}) = \frac{1}{4\hat{X}^2}\sum_{s}\left[(\hat{Y}_{s1} - \hat{Y}_{s2})^2 - 2\hat{R}(\hat{Y}_{s1} - \hat{Y}_{s2})(\hat{X}_{s1} - \hat{X}_{s2}) + \hat{R}^2(\hat{X}_{s1} - \hat{X}_{s2})^2\right]$,

with suffix 1 and suffix 2 denoting sub-samples 1 and 2 respectively.

For the mathematical expressions to compute the RSE of estimates of aggregates and ratios based on the NSS unit-level data, readers may also like to refer to 'Sample Design and Estimation Procedure' featuring in the Appendix to any latest NSS report.

8.2 Precision of Some Key Estimates

8.2.1 *Survey on Household Consumption Expenditure: Household Average MPCE*

The latest estimates of the average MPCE of the households pertain to the Household Consumption Expenditure Survey (HCES): 2022–2023 and HCES: 2023–2024 with survey periods as August to July of the respective years. At the all-India level, the HCES: 2022–2023 covered 8,723 villages and 6,115 UFS blocks with a minimum sample size of 14 villages covered from the UT of Lakshadweep and a minimum of 18 urban blocks surveyed from the State of Goa. In terms of number of total number of sample households, a sample of 155,014 households residing in rural India and 106,732 households in urban India was covered in the HCES: 2022–2023.

The sample size of HCES: 2023–2024 was also of a similar order (with 154,357 rural households and 107,596 urban households surveyed at the all-India level) (see MoSPI, 2022b, 2024c, 2025).

As per the published results of both the surveys, estimates of average MPCE, rural as well as urban, for all-India and for most of the States/UTs are quite robust with RSE below 5% (see Box 8.1). In a few smaller States/UTs,[3] the estimated RSE of average MPCE exceeded 5%. The latest quinquennial survey (NSS 68th round, 2011–2012[4]) on household consumer expenditure with a somewhat smaller sample size (with 7,469 villages and 5,268 urban blocks surveyed at the all-India level) also reveals more or less similar findings.[5]

Box 8.1 RSE of estimated average MPCE of States and UTs

Characteristic	HCES: 2022–2023		HCES: 2023–2024	
	Rural	Urban	Rural	Urban
Estimated RSE: All-India	0.55%	0.64%	0.53%	1.59%
Number of States/UTs by level of RSE:				
RSE being 5% or less	30	31	27	20
5% to 10%	5	5	8	12
More than 10%	1*	0	1^	4@
All	36	36	36	36

Note * Ladakh; ^ Dadra & Nagar Haveli and Daman & Diu; @ Telangana, Chandigarh, Ladakh and Lakshadweep
Source MoSPI (2024c, 2025)

8.2.2 *Periodic Labour Force Survey: LFPR, WPR & UR*

8.2.2.1 Annual Estimates (Rural and Urban)

Based on the PLFS series introduced in 2017, NSSO releases annual estimates of a number of characteristics for rural as well as urban areas and quarterly estimates of key indicators (i.e., LFPR, WPR and UR) for urban areas only. The latest annual estimates relate to the year July 2023–June 2024. This survey (see MoSPI, 2024a)

[3] In the HCES: 2022–23, the States/UTs where RSE of average MPCE exceeds 5% included Delhi, Himachal Pradesh, Andaman & Nicobar Islands, Dadra & Nagar Haveli and Daman & Diu, Ladakh and Puducherry in respect of rural areas and Goa, Sikkim, Uttarakhand, Chandigarh and Ladakh in case of urban areas.

[4] See MoSPI (2014)

[5] In NSS 68th round, the RSE of average MPCE based on the 'Modified Mixed Reference Period' exceeded 5% for Delhi, Goa, Chandigarh, Dadra & Nagar Haveli, Daman & Diu and Lakshadweep in rural areas and for Chandigarh, Dadra & Nagar Haveli, Daman & Diu and Lakshadweep in urban areas (Source: MoSPI (2014)).

covered 6,975 villages and 5,768 urban blocks at the all-India level. Box 8.2 summarizes the frequency distribution of States/UTs[6] by the level of RSE of estimated LFPR, WPR and UR based on usual status (principal and subsidiary statuses taken together, i.e., UPSS approach) as per PLFS: 2023–2024. Since the RSEs are estimated figures based on the sample data, a firmer conclusion about the level of precision of these estimates at the State/UT level can possibly be made based on the average RSE of annual estimates of a few years. Corresponding to Box 8.2, a similar frequency distribution based on the average RSE of the estimates for the last four years (2020–2021 to 2023–2024) is presented in Box 8.3. The following major findings emerge from this statement:

(a) For a large majority of the States/UTs, the estimates of UR (rural/urban; male/female/person) are having quite high margins of error with the average RSE exceeding 10 per cent. The average RSE of UR exceeds even 20% for quite a few States/UTs. This essentially implies that the sample design—particularly the method of stratification and sample size—might require a re-examination for further improvement/increase in the sample size so as to have the State-level estimates of UR within tolerable margins of error.
(b) As regards the males, estimates of LFPR and WPR for both rural and urban areas are robust (i.e., RSE within 5%) for most of the States/UTs. The average RSE of male-WPR lies between 5 and 10% in some cases (mostly in the UTs); and
(c) For females, there are 7 States/UTs in each of rural and urban areas with the RSE of WPR exceeding 10%.

Box 8.2 Frequency distribution of the States/UTs by the level of RSE of LFPR, WPR and UR (UPSS) as per PLFS 2023–24

Category	Number of States/UTs								
	LFPR			WPR			UR		
	RSE: Within 5%	RSE: 5%–10%	RSE: More than 10%	RSE: Within 5%	RSE: 5%–10%	RSE: More than 10%	RSE: Within 5%	RSE: 5%–10%	RSE: More than 10%
R-M	31	3	1	29	4	2	1	0	34
R-F	26	5	4	25	6	4	2	0	33
R-P	31	3	1	29	4	2	0	4	31
U-M	32	4	0	33	3	0	1	10	25
U-F	13	15	8	11	17	8	2	7	27
U-P	33	3	0	33	3	0	4	10	22

Note R: Rural; U: Urban; M: Male; F: Female; P: Person
Source Annual Report, PLFS 2023–2024 (MoSPI, 2024a)

[6] There are 36 States/UTs in the country. However, for rural areas, the UT of Chandigarh is not taken into consideration since there was no rural component for the UT during PLFS: 2023–24.

Box 8.3 Frequency distribution of the States/UTs by the level of average RSE of LFPR, WPR and UR based on the latest four surveys: PLFS 2020–2021 to 2023–2024

Category	Number of States/UTs								
	LFPR			WPR			UR		
	RSE: Within 5%	RSE: 5%–10%	RSE: More than 10%	RSE: Within 5%	RSE: 5%–10%	RSE: More than 10%	RSE: Within 5%	RSE: 5%–10%	RSE: More than 10%
R-M	30	5	0	28	7@	0	0	0	35 (17)
R-F	19	10	6	18	10*	7**	0	0	35 (29)
R-P	29	3	3	28	4	3	0	1	34 (14)
U-M	33	3	0	33	3^	0	0	7	29 (9)
U-F	11	19	6	12	17#	7$	0	1	35 (17)
U-P	35	1	0	34	2	0	0	13	23 (6)

Note R: Rural; U: Urban; M: Male; F: Female; P: Person

Figures within brackets in the last column denote the number of States/UTs for which the average RSE of UR exceeds 20%

@ A & N Islands; Lakshadweep; Puducherry; Goa; Delhi; Dadra & Nagar Haveli and Daman & Diu; Ladakh

* Punjab; Arunachal Pradesh; Tripura; A & N Islands; Nagaland; Manipur; Bihar; Uttarakhand; Mizoram; Haryana

** Puducherry; Dadra & Nagar Haveli and Daman & Diu; Goa; Ladakh; Assam; Lakshadweep; Delhi

^ A & N Islands; Dadra & Nagar Haveli and Daman & Diu; Ladakh

Madhya Pradesh; Uttar Pradesh; Odisha; Arunachal Pradesh; Ladakh; Tripura; Chhattisgarh; Meghalaya; Assam; Gujarat; Sikkim; Jammu & Kashmir; Rajasthan; Puducherry; Uttarakhand; Goa; Delhi

$ Himachal Pradesh; Haryana; Jharkhand; Bihar; Chandigarh; Lakshadweep; Dadra & Nagar Haveli and Daman & Diu

Source Annual reports, PLFS 2020–21, 2021–22, 2022–23 & 2023–24 (MoSPI, 2022a, 2023a, 2023b, 2024a)

8.2.2.2 Quarterly Estimates (Urban)

Quarterly estimates of the LFPR, WPR and UR as per the current weekly status (CWS) approach are released for 22 States (including the NCT of Delhi) apart from all-India. The quarterly estimates of all three indicators for all-India are quite robust with RSE much lower than 5 per cent. So far as the State-level estimates are concerned (see Box 8.4), as observed in respect of the annual estimates, for most of the States, quarterly estimates of UR have high margins of error with RSE exceeding 10 per cent. Some refinements in the method of stratification/sampling and in the sample size may be necessary for improved estimates of State-Level UR. However, the estimates of LFPR and WPR in respect of males for all the States are quite robust (RSE less than 5 per cent). For females, the State-level estimates of LFPR and WPR for certain States have RSEs greater than 5% and the average RSE of four quarters of July 2023–June 2024 even exceeds 10% in respect of the State of Jharkhand.

Box 8.4 Frequency distribution of the States/UTs (Urban) by the level of RSE of quarterly estimates of LFPR, WPR and UR (BASED on CWS) as Per PLFS 2023–2024

Category	Number of States/UTs								
	LFPR			WPR			UR		
	RSE: Within 5%	RSE: 5%–10%	RSE: More than 10%	RSE: Within 5%	RSE: 5%–10%	RSE: More than 10%	RSE: Within 5%	RSE: 5%–10%	RSE: More than 10%
As per the Quarter of April–June 2024									
Male	21	1	0	21	1	0	0	8	14
Female	9	11	2	8	11	3	1	5	16
Person	22	0	0	22	0	0	2	8	12
Based on the average RSE of four quarters of the year July 2023–June 2024									
Male	22	0	0	22	0	0	0	4	18
Female	12	9	1	6	15	1	0	2	20
Person	22	0	0	22	0	0	0	7	15

Source Quarterly Bulletin; April–June 2024; National Statistical Office; MoSPI (2024b)

8.2.3 Survey on Healthcare: Proportion of Persons Ailing, Proportion Hospitalized and Average Expenditure per Hospitalized Childbirth

The latest surveys on healthcare were carried out as part of NSS 71st and 75th rounds (see MoSPI, 2016, 2020b). While NSS 75th round was a survey of one-year duration (July 2017–June 2018), the fieldwork of NSS 71st round was spread over a period of six months only (January–June 2014). As regards the all-India sample size, NSS 75th round covered 8,077 villages and 6,181 UFS blocks. As against the same, NSS 71st round survey was spread over a sample of 4,577 villages and 3,720 UFS blocks at the all-India level. The minimum sample size at the State/UT level in NSS 71st round was 8 villages (from each of Delhi, Chandigarh, Dadra & Nagar Haveli, Daman & Diu, Lakshadweep & Puducherry) and 8 urban blocks from each UT of Dadra & Nagar Haveli, Daman & Diu and Lakshadweep.

Boxes 8.5 and 8.6 respectively summarize the findings on the level of precision of State-level estimates of three indicators namely the proportion of ailing persons, proportion/percentage of persons hospitalized and average expenditure (medical and non-medical combined) per hospitalised childbirth based on NSS 71st round and NSS 75th round. The level of precision of the said average expenditure is not available in the published report of NSS 71st round. Further, the figures of level of precision/ RSE are available only for 18 major States in the concerned NSS 75th round report.

Box 8.5 RSE of proportion of persons ailing and hospitalized, NSS 71st round, January–June 2014

Characteristic	Rural	Urban
I. Indicator: Proportion of Ailing Persons		
Estimated RSE: All-India	1.84%	2.05%
Number of States/UTs by level of RSE:		
RSE being 5% or less	3	1
5% to 10%	11	10
10% to 20%	15	12
More than 20%	7	13
All	36	36
II. Indicator: Percentage of Persons Hospitalized		
Estimated RSE: All-India	1.57%	1.95%
Number of States/UTs by level of RSE:		
RSE being 5% or less	5	0
5% to 10%	18	14
10% to 20%	8	16
More than 20%	5	6
All	36	36

Source MoSPI (2016); NSS report number 574; Health in India; January–June 2014

Box 8.6 RSE of percentage of ailing persons, hospitalized cases per 100,000 persons and average expenditure per hospitalized childbirth during last 365 days, NSS 75th round, July 2017–June 2018

Characteristic	Rural	Urban
I. Indicator: Percentage of Ailing Persons		
Estimated RSE: All-India	1.8%	1.9%
Number of States/UTs by level of RSE:		
RSE being 5% or less	10	7
5% to 10%	8	9
More than 10%	0	2
All	18	18
II. Indicator: Hospitalized Cases including Childbirth per 100,000 Persons		
Estimated RSE: All-India	1.4%	1.7%
Number of States/UTs by level of RSE:		

(continued)

(continued)

Characteristic	Rural	Urban
RSE being 5% or less	6	3
5% to 10%	9	13
More than 10%	3	2
All	18	18
III. Indicator: Average Expenditure per Hospitalized Childbirth during Last 365 Days		
Estimated RSE: All-India	2.0%	5.1%
Number of States/UTs by level of RSE:		
RSE being 5% or less	1	2
5% to 10%	13	12
More than 10%	4	4
All	18	18

Source MoSPI (2020b); NSS report number 586; Health in India; July 2017–June 2018

It may be seen that the all-India estimates of all three indicators are quite robust. As regards the State-level estimates, NSS 75th round estimates for all three indicators are robust (RSE within 10%) for most of the States. However, the NSS 71st round, which was a survey of six-month duration resulted in an RSE exceeding 10% (and sometimes more than 20%) for both the proportion of ailing persons and hospitalized persons in respect of quite a few States/UTs. Given these findings, it may be desirable to conduct the surveys on healthcare spreading over a duration of one-year instead of six months for improved estimates of key indicators.

8.2.4 Survey on Education: Net Attendance Ratio

The latest survey on education was carried out as part of NSS 75th round (July 2017–June 2018) with 8,097 villages and 6,188 urban blocks surveyed in the central sample at the all-India level. NSS Report Number 585 titled 'Household Social Consumption on Education in India' (see MoSPI, 2020a), inter-alia, presents the RSE of net attendance ratio (NAR) in respect of different levels of current attendance by gender at the State/UT level for 36 States and UTs separately for rural and urban areas as well as both rural and urban combined. Box 8.7 summarizes the findings on the level of precision/RSE of the estimates of NAR. The following are the broad conclusions:

- For primary level, the state-wise estimates of NAR for both boys and girls and for both rural and urban areas are quite robust with the RSE within 5% for most of the States and UTs;
- In case of upper primary, the corresponding RSEs are within 5% for most of the States/UTs in case of the boys residing in rural areas. The RSE of NAR exceeds

5% for most of the States/UTs in respect of girls residing in rural areas and both boys and girls residing in urban areas.

- So far as the secondary, higher secondary and post-higher secondary levels are concerned; the RSE of NAR exceeds 10% for a large number of States/UTs for both boys and girls residing in rural as well as urban areas indicating that the State-level estimates of NAR in the higher levels of education are having high margins of error. Some refinements in the sample design/sample size appear to be necessary to overcome this problem in future surveys by way of ensuring in the sample a much larger number of students pursuing such higher levels of education.

Box 8.7 RSE of net attendance ratio by level of current attendance and gender, NSS 75th round, January 2017–June 2018

Distribution of States/UTs by level of RSE of Net Attendance Ratio					
Level of current attendance	RSE	Rural-Male	Rural-Female	Urban-Male	Urban-Female
Primary	Within 5%	Most of the States/UTs	Most of the States/UTs	Most of the States/UTs	Most of the States/UTs
Upper Primary	Within 5%	Most of the States/UTs	–	–	–
	More than 5%	–	Most of the States/UTs	Most of the States/UTs	Most of the States/UTs
Secondary	More than 5%	Most of the States/UTs	Most of the States/UTs	Most of the States/UTs	Most of the States/UTs
	More than 10%	15	12	21	24
Higher Sec (HS)	More than 5%	Most of the States/UTs	Most of the States/UTs	Most of the States/UTs	All the States/UTs
	More than 10%	26	30	29	31
Post HS	More than 5%	Most of the States/UTs	Most of the States/UTs	All the States/UTs	All the States/UTs
	More than 10%	30	32	29	35

Source MoSPI (2020a); NSS report number 585

8.2.5 Survey of Persons with Disabilities: Proportion of Such Persons

The latest survey of persons with disabilities was conducted in NSS 76th round (July–December 2018; MoSPI (2019)). It covered 5,378 rural FSUs (villages/sub-units) and 3,614 urban FSUs (UFS blocks/sub-units) at all-India level. Both rural and urban sample sizes of FSUs was 14 or less in the State of Goa and in all the UTs except Puducherry where urban sample size was 20 and rural sample size was 10. From Box 8.8 indicating the level of RSE of the estimated percentage of persons with disability, it may be seen that for most of the States/UTs, the RSE of estimated proportion of persons with disabilities was within 10%. However, there are quite a few States/UTs having a much higher level of RSE (more than 10%), with the smaller States and UTs dominating this list. One possible remedy to overcome this problem and to have further improved State-level estimates of the proportion of disabled—and for that matter the proportion of disabled by type of disability—is to expand the periodicity of the survey to one year (i.e., essentially doubling the sample size) in place of six months.

Box 8.8 RSE of percentage of persons with disability, July–December 2018

Characteristic	Rural	Urban
Estimated RSE: All-India	0.82%	1.47%
Number of States/UTs by level of RSE:		
RSE being 5% or less	15	2
5% to 10%	12	22
10% to 20%	9	8
More than 20%	0	4
All	36	36

Source MoSPI (2019); NSS report number 583; Persons with disabilities in India

8.2.6 Survey on Situation Assessment of Agricultural Households: Proportion of Agricultural Households, their Average Monthly Income and Average Area Operated per Holding

NSS 77th round (January–December 2019) was the latest round that covered agricultural households (rural only) spread over a random sample of 5,940 villages/sub-units as the FSUs at the all-India level (see MoSPI, 2021a). The number of sample FSUs was small, i.e., 10 or less in 7 States/UTs, i.e., Delhi, Goa, Chandigarh, Dadra & Nagar Haveli, Daman & Diu, Lakshadweep and Puducherry.

Based on this survey, NSS Report Number 587 (MoSPI, 2021a), inter-alia, gives the RSEs of the estimated number of agricultural households, their average monthly income and an average area operated per holding by such households for 28 States (excluding Goa), Group of North-Eastern States and Group of UTs. It may be seen from the report that the RSE of the estimated number of agricultural households is less than 5% for most of the States. For 8 States, the RSE lies between 5 and 10%. In other words, State-wise estimated numbers of agricultural households as per the survey are quite robust/fairly reliable. However, for the group of UTs, the RSE was on the higher side (13.9%).

The report presents RSEs of the estimate of the average monthly income of agricultural households for three reference periods of the agricultural year 2018–2019 namely July–December 2018; January–June 2019 and July 2018–June 2019. It is seen that for 12 out of 28 major States, the RSE of average monthly income exceeded 10% for at least one of the three reference periods.

As regards the average area operated during the agricultural year 2018–2019 per holding by the agricultural households, the State-level estimates are quite robust/ fairly reliable with the RSE within 5% for 10 States and between 5 and 10% for 13 States out of 28 States. For the group of UTs, the estimate has a higher RSE of 12.7 per cent.

8.2.7 Survey on Debt and Investment: Incidence of Indebtedness

The subject of debt and investment also formed part of the enquiry of NSS 77th round and the data were collected from a random sample of 5,940 rural FSUs (villages/ sub-units) and 3,995 urban FSUs (i.e., UFS blocks / sub-units). From the concerned NSS Report Number 588 (see MoSPI, 2021b) it is seen that out of 18 major States for which estimates of RSE are presented,[7] the RSE of the incidence of indebtedness among the households exceeds 10% for 14 States in respect of rural areas and for 4 States in case of urban areas. Appropriate corrective measures are necessary for improving the State-level estimates of the incidence of indebtedness.

8.2.8 Survey of Unincorporated Non-Agricultural Enterprises: Estimated Number of Establishments/ Workers and Annual GVA per Establishment/Worker

In the past, the NSS used to cover enterprises roughly once in five years. However, to meet the growing needs for more frequent data pertaining to enterprises in the unorganized sector, the NSSO introduced the Annual Survey of Unincorporated Sector Enterprises (ASUSE) covering manufacturing, trade and other services with the first such survey carried out during April 2021–March 2022. Due to the coronavirus pandemic, the next/second survey was conducted during October 2022–September 2023 after a gap of few months. The survey was carried out in a sample of 8,495 FSUs in rural areas and 7,887 FSUs in urban areas. Altogether, the data were collected

[7] See Statement R1, Appendix B of NSS report number 588 titled All India debt & investment survey-2019 (MoSPI 2021b).

from a sample of 458,938 establishments, i.e., 258,296 from rural India and 200,642 establishments from urban India.

Box 8.9 summarizes the findings on the level of precision, i.e., RSE of State/UT level estimates of four important indicators, namely (i) total number of establishments covering all the three sectors of manufacturing, trade and other services, (ii) total number of workers engaged by such establishments, (iii) annual GVA per establishment and (iv) annual GVA per worker based on ASUSE 2022–23. The all-India estimates of each of these indicators are quite robust with RSE much below 5%. Also, the overall (i.e., for all the three sectors taken together) State-level estimates of GVA per worker for both rural and urban are fairly reliable (i.e., RSE within 10%) for the majority of the States. However, the RSE of the estimated number of establishments, estimated number of workers and estimated GVA per establishment—even for all industries/sectors combined—exceeds 10% for a large number of States/UTs. This seems to be a major limitation of the survey.

Box 8.9 RSE of estimated number of establishments/workers and estimated annual GVA per establishment/worker, ASUSE: 2022–2023

Characteristic	Estimated Number of Establishments		Estimated Number of Workers		Annual GVA per Establishment		Annual GVA per Worker	
	Rural	Urban	Rural	Urban	Rural	Urban	Rural	Urban
RSE: All-India	3.36%	2.73%	2.94%	2.20%	1.82%	2.89%	1.43%	2.35%
Number of States/UTs by level of RSE:								
5% or less	0	0	0	0	3	1	14	13
5%–10%	4	6	8	13	21	12	16	12
10%–20%	19	14	16	11	7	14	2	8
>20%	12	16	11	12	4	9	3	3
All	35	36	35	36	35	36	35	36

Source MoSPI (2024d); report of NSSO on Operational and economic characteristics, ASUSE: 2022–23

Note For rural, the report did not generate any RSE for the UT of Lakshadweep since only one FSU was surveyed from the UT

References

MoSPI. (2014). *Level and pattern of consumer expenditure 2011–12, NSS 68th round, July 2011–June 2012, Report number 555*. National Sample Survey Office, Ministry of Statistics and Programme Implementation, Government of India.

MoSPI. (2016). *Health in India, NSS 71st round, January–June 2014, Report number 574*. National Sample Survey Office, Ministry of Statistics and Programme Implementation, Government of India.

MoSPI. (2019). *Persons with disabilities in India, NSS 76th round, July –December 2018, Report number 583.* National Statistical Office, Ministry of Statistics and Programme Implementation, Government of India.

MoSPI. (2020a). *Household social consumption on education in India, NSS 75th round, July 2017–June 2018, Report number 585.* National Statistical Office, Ministry of Statistics and Programme Implementation, Government of India.

MoSPI. (2020b). *Health in India, NSS 75th round, July 2017–June 2018, Report number 586.* National Statistical Office, Ministry of Statistics and Programme Implementation, Government of India.

MoSPI. (2021a). *Situation assessment of agricultural households and land and livestock holdings of households in rural India, 2019, NSS 77th round, January–December 2019, Report number 587.* National Statistical Office, Ministry of Statistics and Programme Implementation, Government of India.

MoSPI. (2021b). *All India debt & investment survey–2019, NSS 77th round, January–December 2019, Report number 588.* National Statistical Office, Ministry of Statistics and Programme Implementation, Government of India.

MoSPI. (2022a). *Annual Report, Periodic labour force survey (PLFS), July 2020–June 2021.* National Statistical Office, Ministry of Statistics and Programme Implementation, Government of India.

MoSPI. (2022b). *Instructions to field staff volume-I, Socio-economic survey, July 2022–June 2023, Household consumption expenditure survey (HCES): 2022–23.* National Statistical Office, Ministry of Statistics and Programme Implementation, Government of India.

MoSPI. (2023a). *Annual report, Periodic labour force survey (PLFS), July 2021–June 2022.* National Sample Survey Office, Ministry of Statistics and Programme Implementation, Government of India.

MoSPI. (2023b). *Annual report, Periodic labour force survey (PLFS), July 2022–June 2023.* National Sample Survey Office, Ministry of Statistics and Programme Implementation, Government of India.

MoSPI. (2024a). *Annual report, Periodic labour force survey (PLFS), July 2023–June 2024.* National Sample Survey Office, Ministry of Statistics and Programme Implementation, Government of India.

MoSPI. (2024b). *Quarterly bulletin, Periodic labour force survey (PLFS), April–June 2024.* National Statistical Office, Ministry of Statistics and Programme Implementation, Government of India.

MoSPI. (2024c). *Survey on household consumption expenditure: 2022–23 (HCES: 2022–23), August 2022–July 2023, Report number 591.* National Sample Survey Office, Ministry of Statistics and Programme Implementation, Government of India.

MoSPI. (2024d). *Operational and economic characteristics, Annual survey of unincorporated sector enterprises (ASUSE): 2022–23, October 2022–September 2023.* National Sample Survey Office, Ministry of Statistics and Programme Implementation, Government of India.

MoSPI (2025). *Survey on household consumption expenditure: 2023–24 (HCES: 2023–24), August 2023–July 2024, Report number 592.* National Statistics Office, Ministry of Statistics and Programme Implementation, Government of India.

Chapter 9
Challenges and Way Forward

As highlighted in Chapter 2, over its 75 years, the NSS has generated a wealth of information on a wide range of subjects. The NSS data have been used extensively for official purposes as well as by planners and policymakers, researchers and various other academics. The NSS is undoubtedly a goldmine of information on various socio-economic characteristics relating to the households/persons residing in India as well as the establishments operating from within the country and how these characteristics have changed over time. As a way forward, it will be a welcome step if based on the experiences gathered through different surveys of the NSS as well as other considerations including the global survey practices, attention is paid to address certain critical areas for further strengthening the database of the NSS.

9.1 Sample Size

The NSS started its field operation with a relatively small sample size of FSUs. A sample of 1,833 villages was considered in the 1st round (October 1950–March 1951) at the all-India level.[1] The small sample size continued up to the 7th round which covered a sample of 960 villages and 444 urban blocks and by the 17th round the sample size increased to 3,888 villages and 2,237 urban blocks (see MoSPI (2001)). The significant increase in the sample size had taken place during the 18th round of the NSS (February 1963–January 1964), which covered 8,472 villages and 4,572 urban blocks at the all-India level as part of the central sample.[2] As of now, a sample size of more or less similar order in the rural areas and a higher sample size of nearly 6,000 to 6,500 urban blocks in the household surveys and around 8,000 urban blocks

[1] The 1st round and the 2nd round of the NSS covered only rural areas of the country.

[2] The 'central sample' comprises the sample FSUs/USUs covered by the NSSO through its regular or contractual manpower/investigators. The matching sample of FSUs/USUs covered by the State/UT Governments participating in the NSS Programme is known as the 'state sample'.

G C Manna, *75 Years of the Indian National Sample Survey*, India Studies in Business and Economics, https://doi.org/10.1007/978-981-96-7320-9_9

in the enterprise surveys is being covered in the recent NSS rounds and other surveys including the PLFS, HCES and ASUSE.

From the discussion on the RSE of State/UT-level key estimates in the previous chapter, it is apparent that although the present all-India sample size with its allocation over States/UTs yields robust/fairly reliable estimates for the larger States in respect of many key characteristics excluding a few, the existing sample size appears to be inadequate for the UTs and for some smaller States as well. For improved State/UT level estimates, it is desirable to augment the present sample size for the UTs and the smaller States where the RSEs for most of the characteristics based on the previous surveys are found to be high.

State-level sample size of FSUs in an NSS round is broadly guided by the availability of manpower which is fairly linked to the size of the State. This procedure with more or less the same State-level sample size in different rounds worked well from the practical point of view as it is difficult to transfer the field investigators from one State to another on account of changes in the sample size over rounds. Such a change is likely to create the problem in data collection in view of different languages spoken in different States and other related issues. However, this method of sample allocation to a State/UT may not quite often match with what the required sample size should actually be if the estimates are required to be within acceptable margins of error. An indication of the same is evident from the discussion on the RSEs. Keeping in view the fact that the State-level RSEs are quite high for some rarer characteristics like the UR, proportion of disabled, net attendance ratios in higher education, and even the proportion of employed among the females in respect of certain States/UTs, it will be a welcome step if the current sample size of the related States and UTs can be augmented appropriately with additional manpower engaged for data collection in such surveys.

9.2 The Issue of Underestimation of Population

The primary focus of the NSS is to generate the key estimates related to the subjects of enquiry in a round. However, a concern about the NSS is that it generally underestimates the population, which becomes available as an auxiliary information from the survey. Accordingly, the data users are often advised/compelled to use the rates and ratios based on the survey and appropriately multiply them with the relevant adjustment factors to derive the current estimates of aggregates.

It is worrying to note that the extent of underestimation of population in the surveys has increased a lot, particularly in respect of urban areas (see Box 9.1) with the estimated urban population of the country as per PLFS: 2023–24 being approximately 28% lower than the corresponding projected population.[3] Various factors could contribute to such an underestimation namely, possible inadequacy

[3] The total estimated urban population of 355 million as per PLFS 2023–24 is even lower than the urban population of the country as per Census 2011 which was 377 million (see MoSPI (2024c).

of the urban sampling frame of FSUs,[4] likely problems associated with the listing exercise for preparing the frame of households within the sample FSUs in a survey, inadequate representation of large-sized households in the sample yielding a lower estimate of the average household size—to name a few. It is high time to identify the factors responsible for such underestimation and put in place the corrective remedial measures.

Earlier, the Bimal Roy Committee, constituted by the MoSPI to look into the issue of underestimation of population in the NSS, found the average household size as per the survey conducted by the NSSO around/after Census 2011 to be lower than that based on the census. The said lower average household size was apparently found to be a significant contributing factor to the underestimate of the population based on the NSS since the estimated total number of households from the survey was in a closer agreement with the number of households based on the census. It is important to mention in this context that for selecting the sample households from each FSU in NSS 78th Round (January 2020–August 2021) on Multiple Indicator Survey, one of the strata of USUs comprised such households which had at least 7 members (i.e., the large-sized households) and the remaining listed households of the FSU were stratified into four more strata. A sample of 2 households from the stratum of large-sized households and 18 households from the other four strata were selected for the detailed enquiry. Interestingly, the estimated urban population of the country based on NSS 78th Round was found to be significantly higher by about 16 million (see NSS Report Number 589) than the urban population based on PLFS[5]: 2020–21. However, the PLFS: 2019–20 had yielded a higher estimate of urban population than that of NSS 78th Round.[6] All these estimates of urban population of the country based on the different surveys namely, PLFS: 2019–20 (340 million), NSS 78th Round: 2020–21 (334 million), PLFS: 2020–21 (318 million) and PLFS: 2023–24 (355 million) conducted recently (i.e., much after the census) are lower than the Census 2011 urban population (377 million).

[4] For ensuring the completeness of the UFS frame, i.e., the list of UFS blocks within each town/city, concurrent checks may be put in place to ensure that the entire geographical area of the town/city is covered within the UFS and the union of geographical areas/maps covered by all the UFS blocks within a town/city resembles the true geographical area of the town.

[5] PLFS stratifies the households based on the number of educated members (with general educational level being secondary or above) in the household.

[6] Thus, a single intervention like the formation of a separate stratum of large-sized households for ensuring adequate representation of such households cannot alone resolve the problem of underestimation of population in the NSS and the other surveys of the NSSO. See MoSPI (2021) for PLFS: 2019–20, MoSPI (2022) for PLFS: 2020–21.

Box 9.1 Divergence of estimated population from the projected population

NSS Round/ Survey	Estimated Population ('000)*		Projected Population** ('000)			Extent of Divergence (%)	
	Rural	Urban	Reference Date	Rural	Urban	Rural	Urban
NSS 50	585,022	193,276	Jan 1, 1994	660,971	237,022	11.5	18.5
NSS 55	686,881	234,107	Jan 1, 2000	729,326	278,127	5.8	15.8
NSS 68	774,430	313,826	Jan 1, 2012	841,343	384,688	8.0	18.4
PLFS'17–18	759,277	314,826	Jan 1, 2018	877,270	440,162	13.5	28.5
PLFS'23–24	861,392	354,502	Jan 1, 2024	904,476	494,161	4.8	28.3

*As per the related reports on the employment–unemployment survey

** Reference date is taken as 1st January in between the survey period of the respective surveys. Projected populations corresponding to NSS 50th/55th/68th rounds are based on linear interpolation of the Population Census figures. Those corresponding to the PLFS are also based on linear interpolation using the projected population figures taken from the Report of the technical group on population projections (July 2020), National Commission on Population, Ministry of Health & Family Welfare

Source MoSPI (1996, 2000, 2014a, 2019a, 2024c)

9.3 Surveys on Household Consumption Expenditure

There have been apprehensions from time to time that the NSS fails to ensure adequate representation of rich households in the sample which is likely to affect the estimate of household average MPCE. Another factor that may cause some underestimation of average MPCE is the use of a lengthy schedule/questionnaire adopted for data collection which is likely to cause respondents' fatigue with an effect on the responses, particularly with regard to the non-food items normally finding place towards the end of the questionnaire. To guard against respondents' fatigue, HCES: 2022–23 (MoSPI, 2024d) as well as HCES: 2023–24 (MoSPI, 2025a) split the traditional, i.e., the lengthy questionnaire, used to be canvassed till NSS 68th round (2011–12; see MoSPI (2014b)), into three parts respectively containing (i) food items, (ii) consumables and services items and (iii) items of durable goods with three separate monthly visits to a sample household for completing the interview on each respective part of the questionnaire.[7]

The estimate of private final consumption expenditure (PFCE) as per the National Accounts Statistics (NAS) is a source with which the NSS estimate of aggregate household consumption expenditure (HCE) is often compared. There are differences between the two sources with regard to the coverage and method of estimation of the respective figures. However, a comparison over time with certain adjustments made for the differences in the coverage reveals that the extent of divergence between the NSS estimate of HCE and the PFCE (with the latter source giving the higher estimate)

[7] For further details about the questionnaire, reference periods for different items and method of canvassing the three parts of the questionnaire in the selected households, refer to NSS report number 591 (HCES: 2022–23; see MoSPI (2024d)).

has increased alarmingly to 45.2% during 2011–12 as against the divergence of about 5.5% during the year 1972–73 (see MoSPI (2015b)). Even with the new approach of data collection through the split questionnaires with three visits to a household as followed in the recently conducted HCES, the high order of divergence persists.

In view of the above and also to address the concerns of inadequate representation of affluent households in the survey, some corrective measures may be required to have an improved estimate of HCE. In this effort, apart from the experimentation on the splitting of the questionnaire,[8] it is worth conducting a pilot survey to test the efficacy of a diary-based method of data collection[9] in respect of the low-value and frequently purchased items for improved estimate of the consumption expenditure. Such a method, as adopted in some other countries (see Chapter 6 on global survey practices in this context), is likely to eliminate the problem of recall lapses in the interview-based method of data collection and more so with regard to the items having a 30-day reference period. The adoption of the diary-based method would entail an additional visit to the sample households to hand over the diary and explain to the key respondents the broad concepts of recording the consumption expenditure. To compensate for the additional cost involved in the data collection due to this extra visit to the sample households, if required, the sample size of FSUs can be somewhat reduced. As observed from the findings of certain studies, the reduction of sample FSUs by a certain extent does ensure the generation of the State-level estimates of average MPCE within acceptable margins of error.

Some pilot exercise is also necessary to take a call on appropriate reference periods for the items where the reference periods of 30 days and 365 days are in place. In other words, it will be useful to see whether a shorter reference period for some of these items gives a better result or not. These apart, as regards the sampling frame/method of stratification, in addition to forming first-stage urban strata of affluent area blocks as well as the stratum of relatively affluent households within the sample FSUs, it may be useful to develop a frame of most affluent/super-rich households from the administrative data. Using this frame, a separate dedicated survey of such affluent households with an appropriate sample size can be explored with advance intimation from the government sent to the selected households soliciting their cooperation in the survey. The data as per this list frame-based survey may be used in conjunction with the same from the usual survey (that should exclude the affluent households from the coverage) to arrive at the final aggregate estimate of HCE.

[8] For canvassing the split questionnaires, there can be two approaches namely they can be either canvassed in the same households after a gap of few days (as is the case with the HCES: 2022–23 & 2023–24; see MoSPI (2024d, 2025a) or in three different sets of households with one part of the schedule having detailed food items or consumables and services items or items of durable goods (as per the present splitting of the questionnaire) and abridged versions of the other two parts of the schedules canvassed in a set of households. A properly designed pilot survey will possibly be in a position to suggest the approach which may be more appropriate.

[9] It is needless to mention that there may be difficulty to have diary-based data collection in a small proportion of sample households having no eligible/literate member to record such details. In such households, the existing method of oral enquiry may continue.

9.4 Periodic Labour Force Survey

9.4.1 Improving the State-Level Estimates of UR and Female WPR and Exploring Alternative Stratification Variable for Improved Estimates

The evidence from the last few annual surveys of the PLFS (see Box 8.3 in the previous chapter) clearly brings out the fact that the State/UT level estimates of UR (rural/urban; male/female/person) for most of the States/UTs are not very precise with the RSE of the estimates exceeding 10% and even 20% for quite a few States/ UTs.[10] Similarly, the estimate of female WPR for both rural and urban exceeds 10% for certain States/UTs. Measures are necessary by way of improved stratification, increased sample size, etc. to improve upon the corresponding estimates—particularly the estimates of UR—in the forthcoming surveys. Given the fact that the level of UR is very low which has resulted in high RSE, it may be worth considering whether a separate stratum of households having unemployed members can be formed based on the listing exercise—with possibly oversampling of households from this stratum—to ensure adequate number of such households in the sample which is likely to improve the estimate of UR and its level of precision. In this context, it may be noted that some of the very early rounds adopted such a stratification/procedure. For example, in the 9th round of the NSS (May–November 1955) which happened to be the 1st round covering the subject of employment–unemployment, the listed households in each sample census block were grouped into 2 strata: those consisting of households having at least one unemployed person (first stratum) and the remaining households (second stratum). Thereafter, a sample of 10 households from the first stratum and 6 households from the second stratum was selected by systematic sampling from each block.[11] Similarly, in the 14th round (July 1958–June 1959), for the collection of data on employment–unemployment, all the households in the sample block were arranged with those having unemployed members coming first followed by other households after which a linear systematic sample was drawn. Such a procedure was to ensure proper representation of households having unemployed members in the ultimate sample.

The PLFS—earlier series till December 2024, as well as PLFS 2025—considers the number of members in the household having a level of general education as secondary or above as the stratification variable. Apart from exploring the formation of a separate stratum of households having unemployed members as mentioned in the preceding paragraph, it may be worth exploring whether the remaining households (i.e., those not having unemployed members, which will be in majority) can be

[10] A specified margin of RSE (say r%) implies that the true value in the population may differ from the published/estimated figure by twice the RSE (i.e., 2r %). In other words, if the estimate is having an RSE of 10%, the true value can deviate from the estimate by 20% in either direction. Similarly, if the RSE is 20%, the true value can deviate from the estimate even by 40%.

[11] In the rural areas, such a stratification was not attempted.

stratified based on some alternative variable (other than the number of educated members) to further improve the estimates of LFPR and WPR, particularly the female WPR. A study[12] aimed at estimating the number of persons in the labour force, number of employed and number of unemployed (for rural/urban and for male and female combined) based on the unit-level data of two annual surveys of PLFS 2020–21 and PLFS 2021–22[13] revealed that out of eight auxiliary variables short-listed for this purpose including the present stratification variable being used in the PLFS, the number of adults (age 15 years and above) in the household and also the number of members in the age group of 15–59 years fared much better than the present one as the stratification variable for sampling of households in the PLFS.

9.4.2 Adequately Capturing Female Work Participation

Quite often there are apprehensions about the likely underestimation of female WPR in the employment–unemployment surveys (EUS). However, this concern has not been backed by adequate studies. In the absence of recent alternative data sources other than the PLFS for comparison purpose, an analysis[14] (see Box 9.2.1 & 9.2.2) based on the data of the EUS as per NSS 68th round (July 2011–June 2012), Census 2011, IHDS 2011–12 (November 2011–October 2012) and the second EUS (July 2011–January 2012; see Ministry of Labour & Employment (2012)) as well as the third EUS (October 2012–May 2013; see Ministry of Labour & Employment (2013)) of the Labour Bureau (LB)—with each of these sources adopting the same definition of worker—gives some indication about the likely under-reporting of female workers in the EUS of NSS in view of the fact that each of the other three sources gave a higher estimate of female WPR than that as per NSS 68th round for certain States.

In the above context, it is important to mention that the EUS-NSS including the ongoing PLFS has a direct question on the activity status of the household members, which may miss out some women occasionally engaged in the economic activities from the count of workers unless proper probing is done during the data collection. To minimize this problem and to guard against the possible underestimation of female WPR, it may be useful if the EUS-NSS/PLFS adds a few probing questions in the beginning like whether the women participated during the reference period in agricultural activities including the rearing of livestock (with or without pay) or assisted other family members in the household enterprise (without pay) even on a part-time basis (for at least 30 days in relation to the UPSS approach) so that such women get identified and classified as workers. Some of the countries do ask similar type of questions. For example, as a part of the 'Labor Force Items', the Current Population Survey by the United States Census Bureau has the following question:

---

[12] See Manna and Mukhopadhyay (2023).

[13] See MoSPI (2022) for PLFS: 2020–21, MoSPI (2023a) for PLFS: 2021–22 and MoSPI (2023c) for PLFS: 2022–23.

[14] Findings taken from Manna (2024).

Box 9.2.1 State-wise female WPRs (15 +) as per alternative sources for rural areas (%)

State	EUS-NSS: 2011–12	Census:2011	EUS-LB: 2011–13	IHDS: 2011–12
(1)	(2)	(3)	(4)	(5)
Arunachal Pradesh	40.9	60.8	55.8	55.7
Assam	16.9	34.9	27.8	43.5
Bihar	8.2	32.6	15.3	47.3
Jharkhand	29.4	54.0	36.8	59.6
Karnataka	38.0	51.9	38.4	61.2
Madhya Pradesh	35.8	59.0	38.5	76.3
Nagaland	39.6	64.3	44.2	71.3
Tripura	30.5	36.9	35.0	46.3
All-India	**35.2**	**43.1**	**30.6**	**60.9**

[*Note* Estimates are as per the UPSS approach in the EUS-NSS and EUS-LB, and main and marginal workers are taken together as per Census 2011. The IHDS estimate gives a closer approximation corresponding to the UPSS approach in the NSS (Author's calculation). EUS-LB figures are the simple average of WPR estimates based on the two corresponding surveys. It is worthwhile to mention that, by definition, Census 2011 is likely to give a somewhat higher rate/WPR since the census treated all persons who worked for at least one day during the last year as workers as against the other sources going by a criterion of at least 30 days of work]

'(THE WEEK BEFORE LAST/LAST WEEK), did (name/you) do any unpaid work in the family business or farm?'[15] with response as either 'yes' or 'no'. Similarly, the Labour Force Survey Questionnaire canvassed by the Australian Bureau of Statistics has the following questions: (1) 'Last week, did [you/name] do any work at all in a job, business or farm?' and (2) 'Last week, did [you/name] do any work without pay in a family business?'.[16]

9.4.3 *Addressing Newer Data Requirements*

9.4.3.1 *Measurement of the Size of Gig Economy*

In view of flexibility of work and some other advantages, the gig economy is becoming popular day by day and its size is likely to grow significantly in the coming years. To regulate the social security aspect and put in place other welfare measures for gig workers, the Central Government and State Governments may be interested

[15] Refer to 'Basic CPS items booklet' under the link 'Labor Force Items'; https://www.census.gov/programs-surveys/cps/technical-documentation/questionnaires.html.

[16] See 'Sample labour force survey questionnaire' under the link https://www.abs.gov.au/statistics/research/questionnaires-used-labour-force-survey#the-labour-force-survey-questionnaire.

Box 9.2.2 State-wise female WPRs (15 +) as per alternative sources for urban areas (%)

State	EUS-NSS: 2011–12	Census: 2011	EUS-LB: 2011–13	IHDS: 2011–12
(1)	(2)	(3)	(4)	(5)
Arunachal Pradesh	19.3	30.5	21.1	28.9
Assam	11.6	19.0	19.6	13.1
Bihar	6.7	15.0	7.8	21.6
Jharkhand	8.8	14.0	15.9	19.5
Karnataka	21.9	27.0	25.0	33.1
Mizoram	35.6	42.9	44.1	46.9
Nagaland	18.7	36.6	31.4	48.2
Tripura	14.3	20.2	22.6	31.1
All-India	**19.5**	**20.2**	**16.4**	**23.5**

[Estimates are as per the UPSS approach in the EUS-NSS and EUS-LB, and main and marginal workers are taken together as per Census 2011. The IHDS estimate gives a closer approximation corresponding to the UPSS approach in the NSS (Author's calculation). EUS-LB figures are the simple average of WPR estimates based on the two corresponding surveys. It is worthwhile to mention that, by definition, Census 2011 is likely to give a somewhat higher rate/WPR since the census treated all persons who worked for at least one day during the last year as workers as against the other sources going by a criterion of at least 30 days of work]

in knowing the size of the gig economy.[17] However, based on the available items of information in the PLFS, it is difficult to clearly identify the gig workers from the PLFS data. It will go a long way if the relevant questions/items of information associated with the gig workers are included in the PLFS questionnaire to accurately measure the size of the gig economy and assess the working conditions of the gig workers. Some of the questions which may be pertinent are: whether the workers use the platform for their work or not, whether they are working on a contract and/or piece-rated basis, and whether they have flexibility in work. However, the list of such items may be finalized in consultation with the relevant stakeholders.

9.4.3.2 *Valuation of Women's Unpaid Domestic and Caregiving Services*

As compared to men, a very large proportion of women is engaged in unpaid domestic services and also in the unpaid caregiving services for household members. As per the TUS (2019), 92.3% of the females (age: 15–59 years) were engaged in unpaid domestic services and 32.8% were engaged in unpaid caregiving services for the household members during the year 2019. The corresponding percentages for males were 28.9% and 16.2% respectively. The females also devote a much larger time every day to carry out such services. Among the females in the same age group of 15–59 years, the average time spent per day per participant in the activity was 315 minutes

[17] In fact, the Government of Rajasthan and the Government of Karnataka have come out with the Bill/Draft Bill for the platform-based gig workers.

for unpaid domestic services and 137 minutes for unpaid caregiving services as against the corresponding figures of 95 minutes and 73 minutes respectively for males in the same age group.

The composition of 'work' as per the 19th ICLS Resolution concerning statistics of work recognizes own-use production work of goods and services as distinct forms of work in the new framework for the collection of statistics (see ILO, 2023). While the activities of own-use production work of goods are within the SNA production boundary, the own-use production work of services—which also includes unpaid domestic services and unpaid caregiving services for household members—are activities inside the SNA General production boundary. In view of the same and even otherwise, valuing the above services, particularly performed by women, and what may be its contribution in relation to the GDP should be an important statistical exercise by itself. For facilitating such an exercise, it will be useful if necessary provisions are made in the PLFS questionnaire to identify the regular wage earners pursuing such activities (in others' households) with break-up for the most prominent activities like cooking, cleaning of: utensils/household premises, child care, care for the elderly and so on—so that the PLFS data enables generation of estimates of the average hourly wage of women pursuing such activities. This proxy estimate of average hourly wage can be used in conjunction with the estimates of (a) the proportion of women engaged in various domestic and caregiving activities for the household members and (b) the average number of hours spent per day by women in such activities from the Time Use Survey to value the women's domestic and caregiving services in the unpaid capacity as percentage of GDP.

9.4.3.3 *Capturing Migration Particulars*

The subjects of employment and migration are inter-related in the sense that a significant proportion of the employed/labour force may also be migrants. Accordingly, the inclusion of a few additional items/questions—at least at some regular intervals—on in-migration, out-migration including return migration[18] is likely to enrich the database of the PLFS to a great extent. Some such items which may be of worth consideration are: (i) migration particulars of household members namely whether a migrant, location of the last usual place of residence as well as reason for leaving the usual place of residence and (ii) whether any erstwhile member of the household has migrated out for employment purposes including their present place of stay, activity status and whether they are sending remittances regularly to the parent household.

[18] These types of key information formed part of the NSS 64th round (July 2007–June 2008) survey on E-U including migration. https://www.mospi.gov.in/sites/default/files/publication_reports/531_final1648711999879_0.pdf.

9.4.3.4 *Inclusion of Probing Questions for Those Out of the Labour Force*

India has a considerably lower LFPR/WPR for females as compared to those of many countries. One of the reasons behind the females having low LFPR could be due to their responsibilities to take care of household chores in view of the non-availability of other household members to carry out such activities. However, a proportion of such females may be willing to work if an appropriate type of work is made available with certain facilities provided to them so that the current conditions compelling them to remain out of labour force do not remain as an obstacle to take up employment. It will be useful if certain questions in this regard are also included in the PLFS at least at some regular intervals. In fact, some of the earlier rounds of the EUS had asked some such questions. For example, in the NSS 61st round (July 2004–June 2005), for the persons who attended domestic duties only and for those who attended domestic duties and were also engaged in free collection of goods, sewing, tailoring, weaving, etc. for household use, the following questions were included: (i) whether the person was willing to accept work if work was made available at the household and (ii) if yes: the nature of work acceptable (full time/part time, regular/occasional), type of work acceptable (dairy, poultry, tailoring, ...), whether the person had any skill/experience to undertake that work, and type of assistance required to undertake the work. Such inputs at periodic intervals are likely to provide useful insights that might help the government in formulating proper policies to promote job opportunities, particularly for the females.

9.4.3.5 *Expanding the Database Relating to Labour Underutilization and Statistics of Work*

As part of the current weekly activity particulars, for each of the preceding seven days, if a person was employed on any day, i.e., he or she worked for at least one hour during the day, the PLFS has also a provision to record (a) 'total hours actually worked considering all the work performed during the day' and (b) 'if available for additional work, hours available for such work during the day'. This is valuable information to study the aspect of underemployment situation in the country. However, because of lack of work/working conditions or other reasons, many persons, particularly the casual labourers and the self-employed, may not be working on all the days of the week and instead they might be seeking or available for work, i.e., their activity status may be 'unemployed' on those days. For such days, it may be possible that they were available for work either for the full day or only for some hours during the day. The study on underemployment and labour underutilization will be enriched further if the PLFS also records the number of hours the persons were available for work in respect of those days when their status was 'unemployed'.

Further, in conducting the surveys on E-U, the earlier EUS and the ongoing PLFS go by the concept of 'economic activity' with persons engaged in such economic activities are treated as workers/employed. The economic activities include '(i) production of all goods and services for market (i.e., for pay or profit) including those of government services, (ii) production of primary commodities for own consumption and (iii) own account production of fixed assets'.[19] Three broad activity statuses under which the sample persons get classified in the EUS/PLFS are either 'employed' (as described above), or 'unemployed' (i.e., those not employed but either seeking or available for work) or 'out of labour force' (i.e., those neither employed nor unemployed). It is important to mention in this context that with a view to meeting different objectives, the scope of collection of labour statistics has been expanded through the 19th International Conference of Labour Statisticians (ICLS in 2013) Resolution concerning statistics of work, employment and labour underutilization. The related guidelines recognize the collection of data on five different mutual forms of work for separate measurement: '(a) own-use production work comprising production of goods and services for own final use; (b) employment work comprising work performed for others in exchange for pay or profit; (c) unpaid trainee work comprising work performed for others without pay to acquire workplace experience or skills; (d) volunteer work comprising non-compulsory work performed for others without pay; (e) other work activities (not defined in this resolution)'.[20] Notably, the revised guidelines as mentioned above have a narrower definition of employment which is linked to the work for pay or profit.[21] However, the scope of the revised guidelines for the collection of data relating to labour force statistics through five different forms of work is much wider and it will be a welcome step if the data collection framework of the PLFS can be aligned with the latest ICLS guidelines. Through proper designing of the questionnaire to record various forms of work, it should be possible to derive the estimates of LFPR, WPR and UR as per the recommended guidelines as well as the existing concept for a few years only to maintain a comparable series with the past before finally switching over to the revised guidelines. Some of the countries have already adopted the latest guidelines while many others are yet to do so and this is likely to hinder the cross-country comparison of labour force statistics. For the status of implementation of the revised guidelines, the readers may refer to the document titled 'Quick guide to understanding the impact of the new statistical standards on ILOSTAT databases' of the International Labour Organization (ILO, 2022).

---

[19] The text re-produced from the Annual report, PLFS: 2023–24 (see MoSPI (2024c): Concepts and definitions).

[20] Reproduced from ILO (2023).

[21] Accordingly, the adoption of this definition will lead to a reduction in the employment-to-population ratio.

9.4.4 *Detecting Quarterly Changes in the Population Parameters of WPR and UR for Urban India and Consequent Implications on the Required Sample Size*

The quarterly estimates based on the PLFS are released for the urban areas of 22 States apart from those pertaining to all-India. These quarterly estimates are intended to assess how the labour market situation is changing on a quarterly basis. However, the difference in the State-level quarterly estimates ('current weekly status' approach) of key labour force parameters like the LFPR, WPR and UR as per the PLFS between two quarters is generally found to be not statistically significant[22] (that is, the corresponding confidence belts of the estimates overlap each other) unless there is any abrupt/sharp change in the labour market situation in the intervening period due to occurrence of some events/shocks having a significant impact on the employment situation like the outbreak of COVID pandemic affecting the country as an example. Statements 9.1.1 to 9.1.4 clearly demonstrate the same based on the analysis of estimates pertaining to two consecutive quarters of January–March and April–June for the two years of 2024 and 2020 with the pandemic affecting the quarters for the year 2020. Thus, if there is an abrupt change in the estimate between two quarters, then only the current PLFS methodology appears to be in a position to detect with certainty the direction of change in the State-level estimates and otherwise it fails to do so. Hence, the moot question is: what type of changes the PLFS should make in its sample design including the sample size if it has to equip itself to correctly detect at least the direction of the true quarterly change in the population parameters (like the LFPR, WPR and UR) that may be happening? It will be useful to carry out a comprehensive study in this regard for necessary refinements in the survey methodology. An earlier study[23] (Manna, 2022) based on the quarterly estimates for the year 2020–21 (July–June) concluded that the all-India sample size in the PLFS was more than adequate to detect quarterly changes of 1% point or higher percentage points only in the WPR and UR. As regards the State-level estimates, for most of the States the observed differences between the last two quarters of the year were found to be not statistically significant and the State-level sample size required augmentation even to detect a change of as high as 3% point in the level parameters of WPR and UR.[24]

The earlier study, which was based on the quarterly estimates pertaining to a past year (2020–21), is replicated by taking into consideration the latest available quarterly estimates of WPR and UR till the quarter of July–September 2024. Statement 9.2

---

[22] It means that although the point estimates of LFPR/WPR/UR apparently indicate some increase or decline in the level parameter between the two quarters, the difference is statistically insignificant meaning thereby that the actual direction of change in the true value might be different than what the point estimates seem to suggest or even there could be no such change.

[23] See Manna (2022).

[24] The requirement of sample size to detect a change in a lower percentage point (i.e., smaller than 3% point) is much larger.

Statement 9.1.1 Quarterly change in WPR (all ages) between January–March 2024 and April–June 2024 and their statistical significance

State*	WPR (%)		95% Confidence Limits of the Estimate@			
	January–March 2024	April–June 2024	January–March 2024		April–June 2024	
			Lower Limit	Upper Limit	Lower Limit	Upper Limit
(1)	(2)	(3)	(4)	(5)	(6)	(7)
Andhra Pradesh	38.2	36.0	36.4	40.0	33.7	38.3
Chhattisgarh	39.4	39.5	37.6	41.2	38.0	41.0
Gujarat	40.5	40.7	39.2	41.8	39.1	42.3
Himachal Pradesh	46.1	47.0	44.1	48.1	42.8	51.2
Maharashtra	39.1	39.0	38.0	40.2	38.0	40.0
Uttar Pradesh	32.0	32.7	31.2	32.8	32.0	33.4
All-India	36.9	36.7	36.5	37.3	36.3	37.1

[*Six States with relatively high sample sizes (Maharashtra and Uttar Pradesh), moderate sample sizes (Andhra Pradesh and Gujarat) and relatively low sample sizes (Chhattisgarh and Himachal Pradesh) in the PLFS are considered for the analysis

@ The lower limit and upper limit indicate that the true value of the parameter of interest in the population is likely to lie anywhere between these estimates with a probability of 0.95. The said lower and upper confidence limits (applicable to each of the Statements in this chapter, wherever applicable) are derived by respectively using the formula p(1–1.96r) and p(1 + 1.96r) where ‘p’ denotes the estimate of WPR or UR as the case may be and ‘r’ is the corresponding RSE (without percentage) of the estimate with the values of p and r taken from the Quality Bulletins of the PLFS. The factor 1.96 in the above formulae is the z-score corresponding to a 95% confidence interval

Sources Estimates of WPR are as per MoSPI (2024a) which is the Quarterly Bulletin of PLFS, April–June 2024. The confidence limits of the estimates are as per the author’s calculation]

presents the State-wise quarterly estimates of WPR and UR for persons in the age group of 15 years and above residing in urban India for the latest five quarters till the quarter ending September 2024. A cursory look at the statement will reveal that the current magnitude of quarterly change in the estimates of WPR and UR at the State-level as well as for all-India is much lower than 1% point and it possibly centres somewhere around 0.2 to 0.6 percentage point in most of the cases. Assuming that this type of situation may continue in the immediate subsequent periods, it may be of interest to re-check whether the difference in the current quarterly estimates of PLFS is statistically significant and if not, what may be the implications on the sample size requirement to detect such a change of the current magnitude with a high probability of at least 0.95 assuming that the sample design in the PLFS remains the same.

Statements 9.3.1 and 9.3.2 present the State-level quarterly estimates of WPR and UR respectively for the two quarters, i.e., April–June 2024 and July–September 2024 along with the corresponding 95% lower and upper confidence limits of the

Statement 9.1.2 Quarterly Change in UR (all ages) between January–March 2024 and April–June 2024 and their statistical significance

State	UR (%)		95% Confidence Limits of the Estimate			
	January–March 2024	April–June 2024	January–March 2024		April–June 2024	
			Lower Limit	Upper Limit	Lower Limit	Upper Limit
(1)	(2)	(3)	(4)	(5)	(6)	(7)
Andhra Pradesh	6.3	8.4	5.2	7.4	6.4	10.4
Chhattisgarh	9.2	9.8	6.3	12.1	8.1	11.5
Gujarat	4.6	3.7	3.4	5.8	2.6	4.8
Himachal Pradesh	9.1	9.3	5.0	13.2	7.6	11.0
Maharashtra	7.5	7.3	6.7	8.3	6.5	8.1
Uttar Pradesh	8.3	7.1	6.2	10.4	5.9	8.3
All-India	6.7	6.6	6.3	7.1	6.3	6.9

[*Sources* Estimates of UR are as per MoSPI (2024a) which is the Quarterly Bulletin of PLFS, April–June 2024. The confidence limits of the estimates are as per the author's calculation]

Statement 9.1.3 Quarterly change in WPR (all ages) between January–March 2020 and April–June 2020 and their statistical significance

State	WPR (%)		95% Confidence Limits of the Estimate			
	January–March 2020	April–June 2020	January–March 2020		April–June, 2020	
			Lower Limit	Upper Limit	Lower Limit	Upper Limit
(1)	(2)	(3)	(4)	(5)	(6)	(7)
Andhra Pradesh	34.4	27.8	32.8	36.0	26.0	29.6
Chhattisgarh	35.2	29.7	34.4	36.0	28.2	31.2
Gujarat	38.2	33.3	35.6	40.8	31.6	35.0
Himachal Pradesh	36.4	35.9	33.1	39.7	26.7	45.1
Maharashtra	36.2	23.6	35.4	37.0	22.6	24.6
Uttar Pradesh	29.0	26.2	28.1	29.9	25.1	27.3
All-India	34.1	28.4	33.8	34.4	28.0	28.8

[*Sources* Estimates of WPR are as per MoSPI (2020b) which is the Quarterly Bulletin, PLFS, April–June 2020. The confidence limits of the estimates are as per the author's calculation]

Statement 9.1.4 Quarterly change in UR (all ages) between January–March 2020 and April–June 2020 and their statistical significance

State	UR (%)		95% Confidence Limits of the Estimate			
	January–March 2020	April–June 2020	January–March 2020		April–June 2020	
			Lower Limit	Upper Limit	Lower Limit	Upper Limit
(1)	(2)	(3)	(4)	(5)	(6)	(7)
Andhra Pradesh	12.6	26.0	11.3	13.9	22.6	29.4
Chhattisgarh	9.3	16.1	8.4	10.2	13.4	18.8
Gujarat	3.7	10.5	2.4	5.0	4.7	16.3
Himachal Pradesh	12.8	14.9	7.3	18.3	12.5	17.3
Maharashtra	6.3	35.6	5.5	7.1	32.5	38.7
Uttar Pradesh	9.7	16.7	8.1	11.3	14.9	18.5
All-India	9.1	20.9	8.7	9.5	20.1	21.7

[*Sources* Estimates of UR are as per MoSPI (2020b) which is the Quarterly Bulletin, PLFS, April–June 2020. The confidence limits of the estimates are as per the author's calculation]

respective point estimates.[25] It is of interest to note that the observed differences in the quarterly estimates between the two quarters for both WPR and UR are not statistically significant (meaning an overlap of the two confidence belts) for all the States for which quarterly estimates are released as well as for all-India barring the State of Tamil Nadu in respect of WPR. It indicates the inadequacy of the existing PLFS methodology/sample size to detect the true change in the population—if at all it is there—with certainty. These findings are in conformity with those based on the earlier study. The 95% lower and upper confidence limits of the point estimates as presented in these statements are derived by using the following formulae (see also the footnote of Statements 9.1.1 in this context):

$$y_L = y(1 - 1.96r) \ldots \textit{and} \tag{9.1}$$

$$y_U = y(1 + 1.96r) \ldots \tag{9.2}$$

where y_L denotes the lower confidence limit of y, i.e., the estimate of WPR/UR, y_U denotes the upper confidence limit of y and r represents the estimated RSE of y as published in the Quarterly Bulletin, Periodic Labour Force Survey (PLFS), July–September 2024, National Statistics Office, Ministry of Statistics and Programme Implementation.

[25] The interpretation of such lower and upper confidence limits is that the true value of WPR and UR is likely to lie between these two limits with a probability of 0.95 (i.e., with a chance of 95%).

Statement 9.2 Quarterly estimates of WPR and UR of five consecutive quarters till July–September 2024 as per the CWS approach for all persons (male and female combined) in the age group of 15 years and above for the urban areas

State	WPR (%)					UR (%)				
	Q1	Q2	Q3	Q4	Q5	Q1	Q2	Q3	Q4	Q5
(1)	(2)	(3)	(4)	(5)	(6)	(7)	(8)	(9)	(10)	(11)
Andhra Pradesh	46.7	46.3	47.1	44.4	44.5	7.7	8.0	6.3	8.4	7.3
Assam	48.1	48.9	50.8	50.1	49.0	7.9	7.1	6.2	7.3	7.9
Bihar	40.3	39.7	40.0	38.8	37.6	6.9	6.2	6.6	6.5	7.3
Chhattisgarh	48.2	48.5	48.4	48.6	47.2	10.1	9.6	9.2	9.8	10.4
Delhi	44.4	44.3	43.9	42.0	42.2	3.6	3.3	1.8	2.5	2.6
Gujarat	51.3	51.4	51.0	51.7	51.8	3.1	3.5	4.6	3.7	4.0
Haryana	45.6	46.3	45.9	45.3	45.4	5.2	3.1	4.1	4.7	5.0
Himachal Pradesh	49.2	52.9	57.1	57.6	56.4	14.5	12.3	9.1	9.1	8.7
Jharkhand	38.2	39.6	39.0	38.6	38.5	8.3	7.7	7.4	7.2	6.3
Karnataka	50.0	49.9	49.2	48.5	48.2	4.2	3.6	4.1	4.5	4.8
Kerala	41.7	42.4	42.6	43.3	43.8	9.2	10.3	10.7	10.0	10.1
Madhya Pradesh	45.4	47.0	47.2	47.3	46.9	5.9	6.0	5.4	4.7	6.0
Maharashtra	48.3	48.3	48.8	48.6	49.3	6.9	6.7	7.5	7.3	6.7
Odisha	46.2	46.8	47.0	47.0	45.6	5.9	7.1	7.7	9.2	10.6
Punjab	45.8	48.1	47.8	48.4	49.5	8.8	6.4	7.7	6.7	4.9
Rajasthan	41.4	44.1	45.4	44.8	46.8	12.0	9.8	9.6	10.1	8.9
Tamil Nadu	45.3	45.1	45.3	46.9	49.3	6.1	6.7	6.1	5.5	5.2
Telangana	47.1	48.7	49.4	49.4	50.0	7.7	8.1	8.8	7.1	6.6
Uttarakhand	42.6	42.9	44.0	45.6	44.0	6.1	8.3	8.0	7.2	7.8
Uttar Pr	42.0	42.9	42.7	43.3	42.6	7.0	7.5	8.3	7.1	7.1
West Bengal	48.7	49.3	50.9	49.7	50.8	5.1	5.3	5.1	6.0	5.5
J & K	44.8	45.1	45.8	46.2	46.3	10.7	10.2	11.0	11.5	11.8
All-India	**46.0**	**46.6**	**46.9**	**46.8**	**47.2**	**6.6**	**6.5**	**6.7**	**6.6**	**6.4**

[*Note* Q1: July–September 2023; Q2: October-December 2023; Q3: January-March 2024; Q4: April-June 2024; Q5: July–September 2024)
Source MoSPI (2024b) which is Quarterly Bulletin, PLFS, July–September 2024]

The above finding on the insignificance of the difference in the State-level quarterly estimates leads to the question of what may be the required sample size in the PLFS (n_0) if it needs to detect true quarterly changes of specified magnitudes (Δ) in the population parameter of WPR and UR. To address this question, let the current period estimate of the level parameter of WPR (without percentage) be denoted by p_1(assumed to be the same as the estimate for the quarter of July–September 2024) and the estimate for the immediate next quarter (p_2) by $p_2 = p_1 + \Delta$ (assuming Δ to be positive and the value of p_2 taken to be higher since the current trend in

Statement 9.3.1 Statistical significance of the difference in the quarterly estimates of WPR between April–June 2024 and July–September 2024 for the urban areas (CWS, persons, age: 15 years and above)

State	WPR (%)		95% Confidence Limits (%) of the WPR				Statistical Significance of the Observed Difference
	April–June 2024	July–September 2024	April–June 2024		July–September 2024		
			Lower Limit	Upper Limit	Lower Limit	Upper Limit	
(1)	(2)	(3)	(4)	(5)	(6)	(7)	(8)
Andhra Pradesh	44.4	44.5	41.8	47.0	42.3	46.7	NS
Assam	50.1	49.0	46.9	53.3	45.4	52.6	NS
Bihar	38.8	37.6	36.6	41.0	35.3	39.9	NS
Chhattisgarh	48.6	47.2	47.3	49.9	45.3	49.1	NS
Delhi	42.0	42.2	40.2	43.8	40.1	44.3	NS
Gujarat	51.7	51.8	49.7	53.7	49.4	54.2	NS
Haryana	45.3	45.4	42.9	47.7	43.7	47.1	NS
Himachal Pradesh	57.6	56.4	53.1	62.1	50.5	62.3	NS
Jharkhand	38.6	38.5	36.8	40.4	36.6	40.4	NS
Karnataka	48.5	48.2	46.8	50.2	46.7	49.7	NS
Kerala	43.3	43.8	42.5	44.1	42.4	45.2	NS
Madhya Pradesh	47.3	46.9	45.8	48.8	44.9	48.9	NS
Maharashtra	48.6	49.3	47.5	49.7	48.3	50.3	NS
Odisha	47.0	45.6	44.2	49.8	42.8	48.4	NS
Punjab	48.4	49.5	47.8	49.0	48.0	51.0	NS
Rajasthan	44.8	46.8	41.9	47.7	44.7	48.9	NS
Tamil Nadu	46.9	49.3	45.7	48.1	48.1	50.5	SIG
Telangana	49.4	50.0	47.9	50.9	48.3	51.7	NS
Uttarakhand	45.6	44.0	43.1	48.1	41.5	46.5	NS
Uttar Pradesh	43.3	42.6	42.5	44.1	41.4	43.8	NS
West Bengal	49.7	50.8	48.0	51.4	49.1	52.5	NS
J& K	46.2	46.3	44.4	48.0	44.8	47.8	NS
All-India	**46.8**	**47.2**	**46.3**	**47.3**	**46.8**	**47.6**	**NS**

[*Note* NS: Not Significant; SIG: Significant
Sources Estimates of WPR are as per MoSPI (2024b) which is the Quarterly Bulletin, PLFS, July–September 2024. The confidence limits of the estimates are as per the author's calculation]

Statement 9.3.2 Statistical significance of the difference in the quarterly estimates of UR between April–June 2024 and July–September 2024 for the urban areas (CWS, persons, Age: 15 years and above)

State	UR (%)		95% Confidence Limits (%) of the UR				Statistical Significance of the Observed Difference
	April–June 2024	July–September 2024	April–June 2024		July–September 2024		
			Lower Limit	Upper Limit	Lower Limit	Upper Limit	
(1)	(2)	(3)	(4)	(5)	(6)	(7)	(8)
Andhra Pradesh	8.4	7.3	6.4	10.4	6.5	8.1	NS
Assam	7.3	7.9	5.0	9.6	5.3	10.5	NS
Bihar	6.5	7.3	4.6	8.4	5.1	9.5	NS
Chhattisgarh	9.8	10.4	8.1	11.5	7.8	13.0	NS
Delhi	2.5	2.6	1.5	3.5	1.2	4.0	NS
Gujarat	3.7	4.0	2.6	4.8	2.4	5.6	NS
Haryana	4.7	5.0	3.4	6.0	4.4	5.6	NS
Himachal Pradesh	9.1	8.7	7.6	10.6	6.8	10.6	NS
Jharkhand	7.2	6.3	5.3	9.1	3.6	9.0	NS
Karnataka	4.5	4.8	3.5	5.5	4.0	5.6	NS
Kerala	10.0	10.1	9.2	10.8	8.5	11.7	NS
Madhya Pradesh	4.7	6.0	3.3	6.1	4.8	7.2	NS
Maharashtra	7.3	6.7	6.5	8.1	5.6	7.8	NS
Odisha	9.2	10.6	6.5	11.9	9.0	12.2	NS
Punjab	6.7	4.9	6.0	7.4	3.6	6.2	NS
Rajasthan	10.1	8.9	8.9	11.3	7.8	10.0	NS
Tamil Nadu	5.5	5.2	4.7	6.3	4.4	6.0	NS
Telangana	7.1	6.6	5.1	9.1	5.7	7.5	NS
Uttarakhand	7.2	7.8	4.3	10.1	5.6	10.0	NS
Uttar Pr	7.1	7.1	5.9	8.3	6.0	8.2	NS
West Bengal	6.0	5.5	4.9	7.1	4.0	7.0	NS
J & K	11.5	11.8	10.5	12.5	10.5	13.1	NS
All-India	**6.6**	**6.4**	**6.3**	**6.9**	**6.1**	**6.7**	**NS**

[*Source* Estimates of UR are as per MoSPI (2024b) which is the Quarterly Bulletin, PLFS, July–September 2024. The confidence limits of the estimates are as per the author's calculation]

WPR is on the rise). Then n_0 should be such that the upper confidence limit of the estimate of p_1 should be lower than or equal to the lower confidence limit of p_2 (meaning non-overlapping confidence belts of the respective point estimates). In other words, noting that the product of y and r in Eqs. (9.1) and (9.2) equals to $\sqrt{\frac{Dp_1q_1}{n_0}}$ and $\sqrt{\frac{Dp_2q_2}{n_0}}$, as the case may be,

$$p_1 + 1.96\sqrt{\frac{Dp_1q_1}{n_0}} \le p_2 - 1.96\sqrt{\frac{Dp_2q_2}{n_0}} \tag{9.3}$$

where the notation 'D' in Eq. (9.3) denotes the 'Design Effect'[26] of the survey (PLFS) with $q_1 = 1 - p_1$ and $q_2 = 1 - p_2$.

Accordingly, the value of n_0 can be derived by solving the following equation:

$$p_1 + 1.96\sqrt{\frac{Dp_1q_1}{n_0}} = p_2 - 1.96\sqrt{\frac{Dp_2q_2}{n_0}} \tag{9.4}$$

A little mathematical derivation of equation (9.4) leads to

$$n_0 = \left(\frac{1.96}{\Delta}\right)^2 D\left(\sqrt{p_1q_1} + \sqrt{p_2q_2}\right)^2 \tag{9.5}$$

The value of n_0 for UR is also obtained by using the same Eq. (9.5) with $p_2 = p_1 - \Delta$ (since currently the quarterly UR appears to be on the decline) and the corresponding inequality in Eq. (9.3) considered as:

$$p_1 - 1.96\sqrt{\frac{Dp_1q_1}{n_0}} \ge p_2 + 1.96\sqrt{\frac{Dp_2q_2}{n_0}} \tag{9.6}$$

(signifying that the lower confidence limit of UR at the current quarter is greater than or equal to the upper confidence limit of UR in the next quarter).

The design effect is estimated by using the formula: $D = n.y.r^2/(1 - y)$, with notations 'y' and 'r' (without percentages) being the ones as described in Eqs. (9.1) and (9.2) and the notation 'n' represents the current quarterly sample size of number of persons in the age group of 15 years and above as per the PLFS.

It is of interest to mention that the all-India estimate of D as per the quarters of July–September 2023 and July–September 2024 (ignoring the three-quarters in-between) works out to be 2.86 and 1.93 respectively in respect of WPR (for persons

[26] Design effect of PLFS (or for any survey) is the ratio of variance of an estimator (like the estimate of WPR or UR in the instant case) as per the stratified multi-stage design adopted in the survey to the variance of the estimator under the SRSWOR (unstratified) sampling procedure. Generally, the design effect under a multi-stage design is found to be greater than 1 because of the clustering effect on the ultimate sample.

in the age group of 15 years and above) and as high as 5.94 and 6.22 respectively with regard to UR for persons in the same age group. Since the required sample size (n_0) will increase with the increase in the value D [see Eq. (9.5) in this context] and also because the observed values of D as mentioned above are only estimates based on the sample data, this exercise assumes a modest/smaller value of D as 1.5 for the purpose of calculating the required sample size n_0.

Box 9.3.1 summarizes the findings on the extent of incremental sample size in terms of the ratio of n_0/n for all-India while Box 9.3.2 presents the corresponding findings for the State-level estimates. It is important to note that so far as all-India quarterly estimates are concerned, the present sample size is adequate to detect a change of 0.7 percentage point or higher in the quarterly WPR and a change of at least 0.4 percentage point in the UR. For detecting a quarterly change of lower magnitude in the population parameter which is being frequently observed now, even the existing all-India sample size appears to be inadequate and it requires augmentation in case the sample design remains unaltered. To illustrate, if the quarterly WPR increases by 0.7 percentage point (i.e., from the current level of 47.2% to 47.9%), the existing all-India sample size is adequate to detect this change. In fact, about 87% of the present sample size is expected to detect this change. On the contrary, if WPR increases by a lower percentage point, say by 0.6 percentage point from the current level of 47.2% to 47.8%, the present all-India sample size needs to be about 19% higher than the present sample size to detect this change. Interestingly, these findings on the required incremental sample size for all-India (and also the State-level estimates as discussed in the next paragraph and Box 9.3.2) in terms of the ratio n_0/n do not change significantly if the current level (p_1) of WPR is considered to be higher like 50% or even 60%. It is also interesting to note that to detect a change of similar magnitude, the requirement of sample size is much larger (a little more than 4 times) in case of WPR as compared to the UR.

As regards the requirement of State-level sample sizes, the current sample sizes namely the number of sample persons in the age group of 15 years and above appear to be sufficient for most of the States to detect a quarterly change of as high as 4% point or higher in the WPR and a change of at least 2% point in the UR (Box 9.3.2). For the detection of a quarterly change of smaller percentage points (which appears to be prevailing now), the extent of increment required in the current sample sizes is

Box 9.3.1 Ratio of required sample size to the present sample size for detection of change in the level parameters of specified magnitudes at the all-India level

Indicator: WPR

Δ	0.002	0.003	0.004	0.005	0.006	0.007	0.01
n_0/n	10.67	4.74	2.67	1.71	1.19	0.87	0.43

Indicator: UR

Δ	0.002	0.003	0.004	0.005	0.006	0.007	0.01
n_0/n	2.53	1.11	0.62	0.40	0.27	0.20	0.09

[*Source* Author's calculation]

unduly large. It is of interest to mention that the above findings on the incremental sample size for all-India and States do not change significantly if p_2 is taken as $(p_1 - \Delta)$ for the WPR (i.e., a decline in the WPR) and as $(p_1 + \Delta)$ for the UR signifying an increase in the corresponding level of UR. It is also of interest to note that for both WPR and UR, the minimum value of n_0/n relates to the State of Maharashtra which has the highest quarterly sample size in the PLFS and the maximum value of n_0/n corresponds to the State of Himachal Pradesh having the least sample size in the PLFS. In other words, for the States having smaller sample sizes, the extent of increase in the required sample size is much larger.

Finally, to conclude the discussion, the quarterly estimates of WPR and UR for the States as well as all-India are generally found to be not differing significantly over the quarters meaning thereby that either the parameters of interest (i.e., the trues values in the universe) are not changing so frequently or even if there is some change across the quarters which appears to be small, the existing PLFS methodology seems to be not equipped to accurately detect such small changes with a high probability. The current PLFS quarterly sample sizes at the State level appear to be far short of the required sample size to detect the small quarterly changes in the population parameters of WPR and UR as is being apparently observed now. Even the all-India quarterly sample size appears to be somewhat inadequate to detect a quarterly change of WPR and UR below a certain percentage point as discussed. The above findings are likely to lead to the question of relevance of the high-frequency surveys (quarterly/monthly) unless corrective methodological refinements are incorporated in the survey design to resolve the current deficiencies. A combination of steps like some alternative method of sampling, the adoption of better stratification procedures, enhanced sample size, and improved method of estimation (e.g., the composite method of estimation) might partially address the inadequacy of the quarterly PLFS as discussed above (Manna, 2022).

Box 9.3.2 Ratio of required sample size to the present sample size for detection of change in the level parameters of specified magnitudes at the state level

Indicator: WPR

Δ	0.01	0.02	0.03	0.04
n_0/n—Minimum	4.48	1.12	0.50	0.28
n_0/n—Maximum	59.06	14.72	6.52	3.65

(For $\Delta = 0.03$, n_0/n is greater than 1 for most of the 22 States under study while this ratio is less than 1 for most of the States in respect of $\Delta = 0.04$.)

Indicator: UR

Δ	0.004	0.006	0.01	0.02
n_0/n—Minimum	6.81	2.98	1.04	0.24
n_0/n—Maximum	117.06	51.46	18.11	4.26

(The ratio n_0/n is less than 1 for most of the States in respect of $\Delta = 0.02$.)

[*Source* Author's calculation]

9.4.5 *Use of Composite Estimator for Generating Quarterly Estimates*

Quarterly estimates of LFPR, WPR and UR as per the Current Weekly Status (CWS) by gender and also cross-classified by some specified age groups are being released through Quarterly Bulletins of the PLFS starting with the quarter of April–June 2018 in respect of 22 major States (including the NCR of Delhi) and all-India for the urban areas. These quarterly estimates are prepared as the design-based estimates making use of the data of 4 panels of FSUs and households surveyed in a given quarter under reference. It is important to mention in this context that out of these 4 panels, one panel is introduced afresh during the quarter while the other 3 panels have been surveyed in some of the preceding quarters as well. It will be useful to develop the composite estimators of LFPR, WPR and UR for any quarter by also making use of the data of the other 3 panels surveyed in the previous quarters taking full advantage of the rotational design for improved estimates.[27] Some studies based on the PLFS data show certain gains in efficiency of the composite estimators for quarterly estimates in terms of reduction in the RSE of estimates. It may be useful to take some initiative in this regard.

9.5 Enterprise Survey

9.5.1 *Issues with the Level of Precision of Estimates*

9.5.1.1 *RSE of Estimates as Per ASUSE*

As summarized in Box 8.9 of the preceding chapter, although the State-level estimates of GVA per worker for all industries/sectors combined (i.e., manufacturing, trade and other services) based on the ASUSE 2022–23 are fairly reliable for majority of the States, the same is not the case with estimated total number of establishments, total number of workers and GVA per establishment with the corresponding RSEs exceeding 10% for a large number of States/UTs. The RSE even exceeds 20% in quite a few cases. The problem is likely to be more with the corresponding estimates by State/UT x industry/sector. This seems to be a major limitation of the survey.

[27] Such a practice of deriving composite estimates is adopted by certain countries (like, e.g., Australia and the United States).

9.5.1.2 *Comparative Position of ASUSE with NSS 73rd Round and NSS 67th Round*

It may be of interest to know whether the said problem of high RSE of estimates was existent in the previous enterprise surveys also. Before the ASUSE was introduced as an annual survey, the previous surveys of unincorporated enterprises with a similar coverage in terms of industries were conducted in NSS 73rd round (2015–16) and NSS 67th round (2010–11). There have been some significant changes in the sample design between NSS 73rd round and the ASUSE.[28] Three major changes include: one, the adoption of NSS region as a stratum in the ASUSE as against the district in NSS 73rd round; two, use of the list of UFS blocks as the sampling frame for all cities/towns in the ASUSE against the list of urban CEBs based on the sixth EC as the frame for most of the towns in NSS 73rd round; and three, selection of FSUs by SRSWOR in the ASUSE instead of the PPS sampling procedure generally adopted for selecting villages/urban blocks in NSS 73rd round. The number of FSUs covered in NSS 73rd round (July 2015–June 2016) was almost similar (Rural: 8,484 & Urban: 7,839) to that of ASUSE: 2022–23 (rural: 8,495 & urban: 7,887). But the number of enterprises surveyed in NSS 73rd round was much smaller, i.e., 290,113 (Rural India: 143,179 enterprises & Urban India: 146,934 enterprises) than 458,938 enterprises (rural: 258,296 & urban: 200,642) surveyed in ASUSE 2022–23.

With the availability of results based on ASUSE: 2023–24, Box 9.4 presents the frequency distribution of total number of States/UTs by the level of RSE for three important characteristics as per ASUSE: 2023–24 and ASUSE: 2022–23 vis-à-vis NSS 73rd and NSS 67th rounds. It is interesting to note that in respect of GVA per worker, the total number of States/UTs with RSE within 10% is close as per all the four surveys. However, with regard to estimated numbers of establishments and workers, both NSS 73rd round and NSS 67th round appear to have a distinct edge over ASUSE with the former two surveys showing more States/UTs having the RSE of the estimate within 10 per cent. The higher levels of RSEs of State-level estimates of number of workers/establishments in the ASUSE might have partially contributed to some of the fluctuations observed in the annual estimates of a number of workers as per the ASUSE (see Box 9.5 for certain examples). Given these findings, it may be useful to carry out a deeper study for identifying the factors which might have had some adverse effect on the estimates in the ASUSE and taking steps for further improvement of the estimates in the forthcoming surveys. This may be all the more necessary in view of the annual frequency of the survey.

[28] See Chapter 3 for the details of sample design adopted in NSS 73rd round and ASUSE: 2021–22. The ASUSE: 2022–23 (MoSPI, 2024f) also adopted the same design as that of ASUSE: 2021–22 (MoSPI, 2024e).

Box 9.4 Frequency distribution of States/UTs by level of RSE of key estimates for manufacturing, trade and other services combined; ASUSE: 2023–24, ASUSE: 2022–23 vis-à-vis NSS 73rd round and NSS 67th round

Source	Number of States/UTs by the level of RSE of the estimate				
	Within 5%	5%–10%	10%–20%	More than 20%	All
Estimated Number of Establishments: Rural					
ASUSE: 2023–24	0	3	20	12	35^
ASUSE: 2022–23	0	4	19	12	35*
NSS 73rd Round	8	15	6	7	36
NSS 67th Round	8	16	9	2	35
Estimated Number of Establishments: Urban					
ASUSE: 2023–24	0	7	16	13	36
ASUSE: 2022–23	0	6	14	16	36
NSS 73rd Round	3	16	10	7	36
NSS 67th Round	2	15	9	9	35
Estimated Number of Workers: Rural					
ASUSE: 2023–24	0	7	19	9	35^
ASUSE: 2022–23	0	8	16	11	35*
NSS 73rd Round	4	20	5	7	36
NSS 67th Round	10	11	9	5	35
Estimated Number of Workers: Urban					
ASUSE: 2023–24	0	14	12	10	36
ASUSE: 2022–23	0	13	11	12	36
NSS 73rd Round	2	17	11	6	36
NSS 67th Round	2	15	8	10	35
Estimated Annual GVA per Worker for Market Establishments: Rural					
ASUSE: 2023–24	14	14	6	1	35^
ASUSE: 2022–23	14	16	2	3	35*
NSS 73rd Round	12	18	4	2	36
NSS 67th Round	12	12	7	4	35
Estimated Annual GVA per Worker for Market Establishments: Urban					
ASUSE: 2023–24	19	15	1	1	36
ASUSE: 2022–23	13	12	8	3	36
NSS 73rd Round	9	19	5	3	36
NSS 67th Round	8	15	9	3	35

Note ^ The UT of Chandigarh had no rural component in the survey
* For rural (ASUSE: 2022–23), no RSE was generated for the UT of Lakshadweep as only one FSU was surveyed
Sources MoSPI (2025b, 2024f) Reports on ASUSE 2023–24, 2022–23; MoSPI (2018), NSS report number 582, NSS 73rd round; MoSPI (2013), NSS report number 549, NSS 67th round

Box 9.5 Fluctuations in the annual state-level estimates of number of workers as per the ASUSE

State/UT	Estimated Number of Workers, All-Industries Combined ('000)					
	Rural			Urban		
	2021–22	2022–23	2023–24	2021–22	2022–23	2023–24
Arunachal Pr	64	67	61	30	38	38
Assam	937	857	1,148	442	465	554
Bihar	3,105	4,364	4,391	1,216	1,532	1,461
Gujarat	2,069	2,013	2,183	4,747	4,868	7,018
Haryana	736	945	725	1,332	1,224	1,663
Manipur	145	134	136	122	81	132
Meghalaya	198	194	222	44	127	84
Mizoram	11	8	10	14	9	19
Telangana	1,777	1,658	1,798	2,090	2,023	2,500
A & N Is	13	22	21	14	26	16
Ladakh	13	12	20	17	18	12
All-India	47,637	52,366	56,138	50,251	57,260	64,461

[*Source* Reports on ASUSE (MoSPI, 2024e, 2024f, 2025b)]

9.5.1.3 RSE of Estimates of GVA Per Worker by Industry

In the compilation of GVA for the 'household sector' as per the national accounts statistics (NAS), the enterprise survey estimates of GVA per worker (GVAPW) at the detailed activity level as per the NAS classification (known as compilation categories[29]) are used. Accordingly, it is desirable that such estimates from the enterprise surveys are within acceptable margins of error. The RSEs of estimated GVAPW by such activity are not available in the report on ASUSE. However, the report on NSS 73rd round provides RSEs of GVAPW for each activity category[30] within a sector (i.e., manufacturing, trade and other services) in respect of market enterprises for all India. Box 9.6 summarizes the major findings.

It may be seen that in respect of 15 out of a total number of 45 activity categories in NSS 73rd round, the corresponding RSE of all-India GVAPW exceeds 10 per

[29] The compilation categories in the NAS with the current base year as 2011–12 were formed by regrouping the economic activities at different levels of the National Industrial Classification (NIC), 2008 (see MoSPI, 2008). The composition of the compilation categories is available as Annexure-II of the publication titled 'Changes in methodology and data sources in the new series of national accounts, Base year 2011–12', published in March 2015 by the Central Statistics Office (see MoSPI, 2015a).

[30] Activity categories in NSS 73rd round were formed based on the NIC Codes of the major activity of the surveyed establishments. Altogether, 25 categories for manufacturing sector, 5 categories for the trade sector and 15 activity categories for the other services sector were formed. Details are available in Table A3, Chapter 1 of NSS report number 582, NSS 73rd round (see MoSPI, 2018).

Box 9.6 Frequency distribution of activity categories by the level of RSE of estimated GVAPW, all-India, rural and urban combined, NSS 73rd round, 2015–16

Sector	Number of activity categories formed in NSS 73rd Round	Overall RSE of estimated GVAPW for the sector (%)	Number of activity categories by the level of RSE of GVAPW		
			Within 5%	5%–10%	More than 10%
Manufacturing	25	3.68	7	7	11
Trade	5	1.57	3	1	1
Other Services	15	2.41	7	5	3

Source MoSPI (2018), Table 56, NSS report number 582, NSS 73rd round

cent. Out of these 15 activity categories, the industry composition (in terms of NIC-2008 code) for 10 categories exactly matches with the NAS classification, i.e., the compilation category. Thus, the estimates of GVAPW at the compilation category level appeared to have suffered from high margins of error (see Box 9.7 for details) in NSS 73rd round at least for certain categories.[31] The problem is likely to be more at the level of State/UT x compilation category. One needs to see whether this problem persists in the ASUSE or not. This is an issue which requires attention for necessary corrective measures in the enterprise surveys by way of improved stratification, increased sample size, etc. to improve upon the estimates of GVAPW at the compilation category level.

9.5.2 *Number of Sectors/Industries to Be Covered in a Survey*

Over a long period of time in the past till NSS 63rd round (2006–07), the enterprises engaged in manufacturing, trade and other services were used to be covered in separate rounds of the NSS. For example, NSS 63rd round covered only the services sector excluding the trade sector. Similarly, NSS 62nd round was devoted to the survey of the unorganized manufacturing sector. Prior to that, the trade sector was covered in NSS 57th round (January–December 1997)—that too confining to own account enterprises (i.e., those without any hired worker) and the enterprises with at least one hired worker but total number of workers being 5 or less. The residual category of trading enterprises which had 6 or more workers including at least one hired worker were separately covered by the CSO during October 1996–September 1997. A departure was made in NSS 67th round (2010–11) which covered all unincorporated non-agricultural enterprises covering manufacturing, trade and other services. The

[31] It is important to mention here that the enterprises not maintaining books of accounts form part of the 'household sector' in the NAS and accordingly, the estimates of GVAPW for such enterprises are used. However, the corresponding RSE of GVAPW is not available in the report.

Box 9.7 Activity categories with RSE of GVAPW exceeding 10%, all-India, NSS 73rd round

Activity category* as per NSS 73rd round	Description of activity	Industry Code (NIC-2008)	No. of establishments surveyed	RSE of estimated GVAPW for the activity category (%)
Manufacturing				
M3	Manufacture of beverages	11	895	12.43
M9	Manufacture of paper and paper products	17	572	10.32
M11	Manufacture of coke and refined petroleum products	19	32	15.92
M12	Manufacture of chemicals and chemical products	20	475	13.64
M13	Manufacture of pharmaceuticals, medicinal chemical and botanical products	21	58	26.49
M14	Manufacture of rubber and plastics products	22	644	14.88
M15	Manufacture of other non-metallic mineral products	23	3,260	11.09
M19	Manufacture of electrical equipment	27	268	12.67
Other Services				
S4	Water transport	50	46	14.96
S5	Warehousing and storage	521	151	24.00

(* Same as the compilation category based on the NAS)
Source MoSPI (2018); Report number 582, NSS 73rd round, July 2015–June 2016, National Sample Survey Office, Government of India

practice continued in NSS 73rd round (2015–16) and also continues in the ASUSE introduced from 2021–22.

It is needless to say that the coverage of many sectors/industries in one survey makes the sample design a complex one. Further, given the constraint of an overall sample size of FSUs in a year and the allocated number of enterprises per FSU, the survey is likely to result in the inadequate number of sample enterprises for certain industrial categories leading to the likely loss of precision of related estimates. In the absence of the information on RSEs of estimates for characteristics like number of establishments, number of workers and GVAPW by State/UT x NIC based on the past surveys of NSS 57th, 62nd and 63rd rounds from the published NSS reports, it has not been possible to present the relative efficiency of these surveys (devoted to one sector only), if any, over the subsequent enterprise surveys of NSS 67th round, NSS 73rd round and ASUSE. As a way forward, it will be useful to carry out such analysis and take appropriate measures for improving the key estimates based on enterprise surveys.

9.5.3 *Separate Survey of Bigger Establishments*

For the surveys on the manufacturing sector, there has been a well-established procedure where the establishments registered under the Factories Act, 1948—being mostly the larger units having a much higher average size in terms of output, employment, etc.—get covered through the Annual Survey of Industries (ASI) and most of the residual units are covered through the NSS/ASUSE. Such a mechanism does not exist at present for non-manufacturing establishments. Some efforts were made in NSS 62nd and 63rd rounds for developing a list frame of larger establishments from where a random sample was drawn for detailed enquiry. This sample was supplemented with the traditional area-frame-based sample of establishments. Given that there is heterogeneity even among the unorganized/unincorporated non-manufacturing establishments in terms of employment and other characteristics, it will be useful to develop a list frame of bigger establishments by tapping different data sources like the Economic Census, EPFO, ESIC, State-level Business Registers and the database of GSTN. The same may be updated regularly for use as a list frame that can be used to draw a representative sample of bigger establishments in the non-manufacturing sector for the purpose of data collection. The residual smaller establishments may as usual be covered through the area-frame-based survey like the ASUSE. The necessary resource implications in view of such a change in the method of data collection need to be addressed for improved estimates of key characteristics from the enterprise surveys. It is worth mentioning in this context that an attempt to cover the larger enterprises in the services sector (excluding the financial sector) through a list frame-based survey was made in NSS 74th round (July 2016–June 2017) as a prelude to the contemplated Annual Survey of Services Sector Enterprises (ASSSE). The survey revealed many interesting findings.[32] Subsequently, the MoSPI has also conducted a pilot study on ASSSE with phase-I conducted during June–August 2024 and phase-II during October 2024–January 2025 to know the characteristics of such enterprises.

[32] For details of the survey methodology and major findings, readers may refer to Technical report TR (74/2.35): Survey on services sector in India, NSS 74th round (MoSPI, 2019b).

9.5.4 Bridging Data Gaps for the Study on Skill Mismatch/ Shortage

The Government of India, with the Ministry of Skill Development and Entrepreneurship as the nodal ministry, have initiated a number of short-term and long-term training schemes as well as other schemes/initiatives to promote skill upgradation of the people belonging to various target groups of the society so that they have better chances of employability. For example, the programme titled 'Skill Acquisition and Knowledge Awareness for Livelihood Promotion (SANKALP)' is a programme aimed at improving 'short-term skill training' to bring in 'better market connectivity and inclusion of marginalized sections of the society'.[33] The PLFS contains certain useful information like the level of general education, level of technical education and formal vocational/technical training particulars of household members and these pieces of information can together be utilized to approximately categorize the size of the present workforce according to the current skill level (like highly skilled, skilled, low skilled and unskilled). However, with the passage of time and more so for the future, the industries might also be looking for employees competent enough to perform various types of jobs depending upon the roles they are required to play under the likely changes in the labour market dynamics. Many of the firms may be currently compelled not to fill up certain positions in view of the non-availability of the manpower having the requisite skills. Inclusion of certain questions relating to job and skill requirements of the enterprises (as well as on the vacancy positions by reasons for such vacancies) in the enterprise surveys (ASI/ASSSE/ASUSE)[34]—particularly in respect of the bigger enterprises—is likely to be valuable inputs for conducting the skill training programmes. Further, this information together with the information available from the PLFS as described above may provide useful insights for the studies relating to skill mismatch or skill shortages.

9.6 Survey on Income

The NSS covered the subjects of income and expenditure during its 9th to 14th rounds. Thereafter, it has avoided collecting the data on income. Subsequently, the Pilot Survey of Income, Consumption and Savings (Part I and II)[35] came out with certain findings on the quality of data on income based on the pilot survey. Reported data on

---

[33] Source: Ministry of Skill Development and Entrepreneurship, Government of India.

[34] Another possible source to collect such information is the Occupational Wage Surveys (OWS) conducted earlier by the Labour Bureau, Ministry of Labour and Employment, Government of India. The last/7th round of OWS was conducted during July, 2016–December, 2017 covering 56 selected industries under different sectors (45 in Manufacturing, 4 in Mining, 3 in Plantation and 4 in Services Sector).

[35] Part 1 of the survey was published as Report number 390 (see Department of Statistics (1993)) and Part II as Report number 398 (see Department of Statistics (1995)).

income (and also savings) was not found to be of high quality. It was also observed that the schedules of enquiry did not 'seem to have been well designed' to collect data on non-monetized components of income and consumption. The pilot study recommended further full-scale pilot surveys to arrive at a satisfactory methodology for household income surveys. Income being an important subject of enquiry and given the fact that the NSS 59th, 70th and 77th rounds have already collected data on income relating to farmer/agricultural households and the PLFS has been collecting data on remuneration/earnings of the self-employed, regular wage earners and casual labourers, it is high time that the NSS explores conducting a survey on household income taking lessons from the pilot survey and the PLFS. However, before doing so, a pilot survey with an adequate sample size and a reasonable geographical coverage may be desirable.

9.7 Need for Methodological Studies

The NSS started its journey way back in 1950. Over the years there have been occasional major changes in the survey methodology including the sample design and method of data collection. Many such interventions appear to have been made without adequate studies or prior pilot surveys. Apart from considering some of the suggestions already given in this chapter, it will be useful to conduct methodological studies in the following areas (illustrative only) with findings taken into consideration for refinement of survey instruments if necessary:

- Assessing relative efficiency of circular systematic sampling of FSUs (retaining geographical arrangement of villages/towns/urban blocks as per the sampling frame) with PPS/equal probability as prevalent in many NSS rounds till NSS 57th round (see Box 2.2, Chapter 2) vis-à-vis the PPSWR/SRSWOR method of sampling[36]
- Efficacy of dividing the large villages into notional sub-units (SUs) and selecting the SUs as the FSUs by SRSWOR in the recent surveys/rounds since NSS 76th round vis-à-vis the PPSWR method of sampling of villages adopted in many NSS rounds
- Relative efficiency of covering a single sector (broad industry) vis-à-vis multiple sectors in the enterprise surveys/ASUSE in terms of RSE of key estimates
- Assessment of the efficacy of the diary-based method of data collection with regard to frequently purchased and low-value consumption items over the interview method to capture the true HCE in a better manner
- Studying the relative efficiency of three methods for more accurately estimating the average MPCE namely: canvassing (a) the three split schedules of HCE used in HCES: 2022–23 & 2024 with three separate visits to the same households as

[36] The method of circular systematic sampling (with PPS or with equal probability), based on the arrangement of all the units in the stratum as per their geographical locations, ensures a wider geographical spread of the Sample FSUs within the stratum.

in the HCES,[37] (b) one of the three split schedules—along with abridged versions of the other two split schedules as applicable—canvassed in one among the three different sets of sample households within the FSU (involving one visit only to a sample household) and (c) the full schedule (i.e., the combination of all the three split schedules) in the same households through a single visit to the household

- Assessing the gain in efficiency of using composite estimator of LFPR, WPR and UR in the Quarterly PLFS (urban) over the current method (i.e., design-based estimator making use of the data of only the sample households surveyed in a given quarter)
- A pilot study to recommend a proper methodology to estimate household income for the possible launching of a survey on this important subject which the NSS has avoided for a long time

9.8 The Issue of Contract Investigators

Of late, apart from many new surveys being taken up by the NSSO like the Time Use Survey (2019 & 2024),[38] Multiple Indicator Survey (NSS 78th Round),[39] Survey on the use of AYUSH (as part of NSS 79th Round)[40] and Comprehensive Annual Modular Survey (as a part of NSS 79th Round)[41] to fulfil the growing demands, the periodicity of surveys on employment–unemployment and enterprises (through PLFS and ASUSE respectively) has been increased as the annual series. This additional workload has led to the engagement of contract investigators for carrying out the primary fieldwork in many surveys which was not the case earlier.

The NSSO organizes training programmes to train the contract investigators on the concepts and definitions used in the surveys. However, quite often concerns are raised about the data quality of the surveys with the fieldwork carried out by the contract investigators. To validate such concerns on a scientific basis, it will be useful if some studies/pilot surveys covering two or three important subjects including the household consumption expenditure, and employment–unemployment can be conducted by using a mix of both regular investigators (who are expected to be more conversant with the concepts and definitions used in the NSS) and contract

---

[37] To quote from the NSS report no. 591: Survey on household consumption expenditure: 2022–23, 'The consumption basket of items was divided into three broad categories, namely, (i) Food items, (ii) Consumables and Services items, and (iii) Durable goods in HCES: 2022–23. Thereafter, three separate questionnaires, namely, Questionnaire: FDQ, Questionnaire: CSQ and Questionnaire: DGQ covering the three categories were designed and canvassed in the selected households in three separate consecutive monthly visits'. See MoSPI (2024d).

[38] See MoSPI (2020a) for TUS 2019, and for TUS 2024 see https://pib.gov.in/PressReleasePage.aspx?PRID=2106113.

[39] See MoSPI (2023b).

[40] https://www.mospi.gov.in/sites/default/files/publication_reports/Fact_sheet-Survey_on-Ayush.pdf.

[41] https://www.mospi.gov.in/sites/default/files/publication_reports/CAMS%20Report_October_N.pdf.

investigators roughly in the ratio of 50:50. The respective two teams may be deployed to carry out the fieldwork in two independent samples of FSUs and USUs drawn on a scientific basis with a reasonable geographical coverage and sample size. Thereafter, the data of the respective two teams may be assessed to take a firmer view about the data quality of the contract investigators and put in place the corrective measures if so warranted.

References

Department of Statistics. (1993). *Report on the pilot study of income, consumption and savings—Part I: Income, Methodological study, September 1983–December 1984, Report number 390.* National Sample Survey Organisation, Government of India.

Department of Statistics. (1995). *Report on the pilot study of income, consumption and savings—Part II, Methodological study, September 1983–December 1984, Report number 398.*, National Sample Survey Organisation, Government of India.

ILO. (2022). *Quick guide to understanding the impact of the new statistical standards on ILOSTAT databases.* Geneva: International Labour Office, Retrieved online at https://www.ilo.org/publications/quick-guide-understanding-impact-new-statistical-standards-ilostat

ILO. (2023). *Resolution II: Resolution to amend the 19th ICLS resolution concerning statistics of work, employment and labour underutilization.* 21st International Conference of Labour Statisticians (Geneva, 11–20 October 2023). Geneva: International Labour Office, Retrieved online at https://www.ilo.org/sites/default/files/wcmsp5/groups/public/@dgreports/@stat/documents/normativeinstrument/wcms_230304.pdf

Manna, G. C. (2022, April 23). On the adequacy of the quarterly PLFS. *Economic & Political Weekly, LVII* (17), 13–17, April 23.

Manna, G. C. (2024). *Measurement issues in estimating female WPR and in valuing their domestic services.* Paper presented in the National Seminar on Women and Employment in India: Issues and Challenges. Jointly organized by the Institute for Human Development and Rajiv Gandhi University, Itanagar during 28–30 August.

Manna, G. C., & Mukhopadhyay, D. (2023, September 2). Exploring alternative stratification variables for sampling of households in the PLFS. *Economic & Political Weekly, LVIII* (35), 16–19.

Ministry of Labour & Employment. (2012). *Report on second annual employment & unemployment survey (2011–12) Volume I.* Labour Bureau, Government of India.

Ministry of Labour & Employment. (2013). *Report on third annual employment & unemployment survey (2012–13) Volume I.* Labour Bureau, Government of India.

MoSPI. (1996). *Employment and Unemployment in India, 1993–1994, Key results, NSS 50th round, July 1993–June 1994, Report number 406.* National Sample Survey Organisation, Ministry of Statistics and Programme Implementation, Government of India.

MoSPI. (2000). *Employment and unemployment in India, 1999–2000, Key results, NSS 55th round, July 1999–June 2000, Report number 455.* National Sample Survey Organisation, Ministry of Statistics and Programme Implementation, Government of India.

MoSPI. (2001). *Concepts and definitions used in NSS.* Golden Jubilee Publication, National Sample Survey Organisation, Ministry of Statistics and Programme Implementation, Government of India.

MoSPI. (2008). *National industrial classification (all economic activities), 2008.* Central Statistical Organisation, Ministry of Statistics and Programme Implementation, Government of India.

MoSPI. (2013). *Economic characteristics of unincorporated non-agricultural enterprises (excluding construction) in India, NSS 67th round, July 2010–June 2011, Report number 549.*

National Sample Survey Office, Ministry of Statistics and Programme Implementation, Ministry of Statistics and Programme Implementation, Government of India.

MoSPI. (2014a). *Employment and unemployment situation in India, NSS 68th round, July 2011–June 2012, Report number 554.* National Sample Survey Office, Ministry of Statistics and Programme Implementation, Government of India.

MoSPI. (2014b). *Level and pattern of consumer expenditure 20ll-12, NSS 68th round, July 2011–June 2012, Report number 555,* National Sample Survey Office, Ministry of Statistics and Programme Implementation, Government of India.

MoSPI. (2015a). *Changes in methodology and data sources in the new series of national accounts, Base year 20ll-12.* Central Statistics Office, Ministry of Statistics and Programme Implementation, Government of India.

MoSPI. (2015b). *Report of the committee on private final consumption expenditure.* Central Statistics Office, Ministry of Statistics and Programme Implementation, Government of India.

MoSPI. (2018). *Economic characteristics of unincorporated non-agricultural enterprises (excluding construction) in India, Report number 582, NSS 73rd round, July 2015–June 2016.* National Sample Survey Office, Ministry of Statistics and Programme Implementation, Government of India.

MoSPI. (2019a). *Annual report, Periodic labour force survey (PLFS), July 2017–June 2018.* National Statistical Office, Ministry of Statistics and Programme Implementation, Government of India.

MoSPI. (2019b). *Technical report on services sector enterprises in India, NSS 74th round, July 2016–June 2017, Technical report No.TR (74/2.35).* National Sample Survey Office, Ministry of Statistics and Programme Implementation, Government of India.

MoSPI. (2020a). *Time use in India–2019, Time use survey, January–December 2019.* National Statistical Office, Ministry of Statistics and Programme Implementation, Government of India.

MoSPI. (2020b). *Quarterly bulletin, Periodic labour force survey (PLFS), April–June 2020.* National Statistical Office, Ministry of Statistics and Programme Implementation, Government of India.

MoSPI. (2021). *Annual report, Periodic labourforce survey (PLFS), July 2019–June 2020.* National Statistical Office, Ministry of Statistics and Programme Implementation, Government of India.

MoSPI. (2022). *Annual report, Periodic labourforce survey (PLFS), July 2020–June 2021.* National Statistical Office, Ministry of Statistics and Programme Implementation, Government of India.

MoSPI. (2023a). *Annual report, Periodic labour force survey (PLFS), July 2021–June 2022.* National Sample Survey Office, Ministry of Statistics and Programme Implementation, Government of India.

MoSPI. (2023b). *Multiple indicator survey in India, NSS 78th round, 2020–2021. NSS report number 589.* National Sample Survey Office, Ministry of Statistics and Programme Implementation, Government of India.

MoSPI. (2023c). A*nnual report, Periodic labour force survey (PLFS), July 2022–June 2023.* National Sample Survey Office, Ministry of Statistics and Programme Implementation, Government of India.

MoSPI. (2024a). *Quarterly bulletin, Periodic labourforce survey (PLFS), April–June 2024.* National Statistical Office, Ministry of Statistics and Programme Implementation, Government of India.

MoSPI. (2024b). *Quarterly bulletin, Periodic labour force survey (PLFS), July–September 2024.* National Statistical Office, Ministry of Statistics and Programme Implementation, Government of India.

MoSPI. (2024c). *Annual report, Periodic labour force survey (PLFS), July 2023–June 2024.* National Sample Survey Office, Ministry of Statistics and Programme Implementation, Government of India.

MoSPI. (2024d). *Survey on household consumption expenditure: 2022–23, (HCES: 2022–23), August 2022–July 2023, Report number 591.* National Sample Survey Office, Ministry of Statistics and Programme Implementation, Government of India.

MoSPI. (2024e). *Operational and economic characteristics, Annual survey of unincorporated sector enterprises (ASUSE): 2021–22, April 2021–March 2022.* National Sample Survey Office, Ministry of Statistics and Programme Implementation, Government of India.

MoSPI. (2024f). *Operational and economic characteristics, Annual survey of unincorporated sector enterprises (ASUSE): 2022–23, October 2022–September 2023.* National Sample Survey Office, Ministry of Statistics and Programme Implementation, Government of India.

MoSPI. (2025a). *Survey on household consumption expenditure: 2023–24, (HCES: 2023–24), August 2023–July 2024, Report number 592.* National Statistics Office, Ministry of Statistics and Programme Implementation, Government of India.

MoSPI. (2025b). *Operational and economic characteristics, Annual survey of unincorporated sector enterprises (ASUSE): 2023–2024, October 2023–September 2024.* National Statistics Office, Ministry of Statistics and Programme Implementation, Government of India.

Correction to: Glimpses of Sample Design

Correction to:
Chapter 3 in: G C Manna, *75 Years of the Indian National Sample Survey*, India Studies in Business and Economics, https://doi.org/10.1007/978-981-96-7320-9_3

In the original version of this chapter, a new paragraph [The change in the approach of data collection through split schedules and multiple visits to a household is likely to yield improved estimates of average MPCE and consequently the poverty rates based on HCES: 2022–23 and HCES: 2023–24 because of the reduction in respondents' fatigue. However, this has led to incomparability of these estimates with those as per the past series. Manna (2024) suggested a remedy to overcome this problem] and a reference [Manna, G. C. (2024). A new methodology with some issues. THE HINDU, Opinion, April 9.] have been included. The chapter has been updated with the changes.

The updated version of this chapter can be found at
https://doi.org/10.1007/978-981-96-7320-9_3

G C Manna, *75 Years of the Indian National Sample Survey*, India Studies in Business and Economics, https://doi.org/10.1007/978-981-96-7320-9_10

MIX
Papier aus verantwortungsvollen Quellen
Paper from responsible sources
FSC® C105338

If you have any concerns about our products,
you can contact us on
ProductSafety@springernature.com

In case Publisher is established outside the EU,
the EU authorized representative is:
Springer Nature Customer Service Center GmbH
Europaplatz 3, 69115 Heidelberg, Germany

Printed by Libri Plureos GmbH
in Hamburg, Germany